A Gentle Introduction to
Knots, Links
and Braids

A Gentle Introduction to
Knots, Links
and Braids

Ruben Aldrovandi
São Paulo State University (UNESP), Brazil

Roldão da Rocha Jr.
Federal University of ABC, Brazil

W **World Scientific**

NEW JERSEY · LONDON · SINGAPORE · BEIJING · SHANGHAI · HONG KONG · TAIPEI · CHENNAI · TOKYO

Published by

World Scientific Publishing Co. Pte. Ltd.
5 Toh Tuck Link, Singapore 596224
USA office: 27 Warren Street, Suite 401-402, Hackensack, NJ 07601
UK office: 57 Shelton Street, Covent Garden, London WC2H 9HE

British Library Cataloguing-in-Publication Data
A catalogue record for this book is available from the British Library.

A GENTLE INTRODUCTION TO KNOTS, LINKS AND BRAIDS

ISBN 978-981-124-848-1 (hardcover)
ISBN 978-981-124-932-7 (paperback)
ISBN 978-981-124-849-8 (ebook for institutions)
ISBN 978-981-124-850-4 (ebook for individuals)

For any available supplementary material, please visit
https://www.worldscientific.com/worldscibooks/10.1142/12591#t=suppl

Desk Editor: Rhaimie Wahap

Typeset by Stallion Press
Email: enquiries@stallionpress.com

Printed in Singapore

to Arqayadustr, my Parents and to the Family
ROLDÃO DA ROCHA JR.

to Paulo Leal Ferreira, in memoriam
RUBEN ALDROVANDI

Contents

Preface

Braids, knots, and links have been infusing, since the antepenultimate century, an increasing interest in the no man's land layer between mathematics and physics[1]. Knot theory is an involved mathematical theory, also related to what is usually called low-dimensional topology. It may seem surprising that a mathematical theory of knots was first developed in 1771 by Vandermonde, who attributed topological features to the geometrical properties of knots [Van71]. It followed after preliminary sketches in the setup of the so-called geometry of position (*geometria situs*), proposed three decades before by Leibniz and Euler. Knots were more formally investigated by Gauss, who introduced in 1833 a mathematical structure, currently known as the Gauss linking integral, for computing the linking number associated with two knots. Gauss supervised a student, Listing, also

---

[1]For a qualitative appraisal, see Ref. [Bir91], and for an involved treatment Ref. [Ati91] is more appropriate.

engaged in the study of knot theory. The 8-knot had been named
after him, being equivalently called the Listing's knot. Around
1867, motivated by experiments and studies about smoke rings[2],
accomplished by Tait, Thomson (also known as Lord Kelvin)
proposed that atoms would be described by knots of whirl vortexes [Tho82, Tho83, Tho85]. This concept asserted that knots
and links formed the very intimate structure that underlies any
chemical element. Helmholtz had inspired Tait, implementing
the foundations of vortex dynamics [DS06]. As a result, the theory describing the dynamics of vortex rings, in inviscid incom-
pressible fluid flows, was then formulated [Hel58, Hel67, Hel69].

Tait, subsequently, listed a comprehensive table of knots, also
formulating the so-called Tait conjectures [LR97]. They assert
that any reduced diagram, having an alternating link, has the
fewest possible crossings, which means that the crossing number
of alternating links[3] is a knot invariant. Besides, the conjectures
also assert that an anti-chiral alternating link has zero writhe and
that given any two reduced alternating diagrams of an oriented
alternating link, one can be transformed to another, by an appropriate sequence of flypes[4]. Tait conjectures have been finally
demonstrated in 1987 [Kau871, Mur87, Thi87]. Maxwell, already
a prominent scientist, and a workmate and long-life friend of
both Tait and Thomson, also developed a strong interest in knot
theory. Maxwell studied Listing's work on knots. A card sent to
Tait in 1867 suggests that Maxwell, motivated by knot applications in electromagnetism, analysed Helmholtz's manuscripts on
vortexes and knots. Maxwell then proposed a physical meaning
to the Gauss linking integral, asserting that an electrical current
induced on a knot yields a magnetic field [DS06]. The linking

---

[2]Smoke tori, indeed.

[3]An alternating knot has a knot diagram wherein crossings alternate between undercrosses and overcrosses.

[4]A flype is a rotation by an angle π of a tangle.

number, on the other hand, was proposed to describe the work done by a charged particle, that moves across the second knot. Besides, Maxwell also investigated smoke ring vortexes, modelling them by three Borromean rings.

The interest in knot theory is also a consequence of a growing awareness of the importance of topological and algebraic properties in many physical phenomena. On the other hand, it emerges from the discovery of a rather unexpected relation between knots and algebraic theory, namely, the operator algebras of von Neumann type. A lot of common ground comes in between, relating non-commutative geometry, arising from those algebras (from the mathematical point of view) to the so-called quantum groups (from the physical viewpoint). The main point of contact is evinced from two apparently unrelated questions in mathematics: the solubility of some lattice models in statistical mechanics and the integrability of – at first, Schrödinger – differential equations, for special problems. Both are encoded in a common algebraic condition, the Yang-Baxter equation [McG64]. Braids constitute highly intuitive groups, which were first studied by Artin [Art47]. Also, braids have an intimate relationship with knots and links. By the way, the reader is encouraged to proceed with real experiments with strings, to get the feeling and the intuition throughout this book.

Topologists in the early part of the 20th century, including for example Dehn and Alexander, studied knots from the knot group and invariants from homology theory, such as the Alexander polynomial for the two trefoil knots [MD14]. Dehn also developed the so-called Dehn surgery, which relates knots to the 3-manifolds theory, also formulating the Dehn problems in the group theory setup, like the word problem. The seminal work of Alexander consisted of a pioneering approach to topological invariants of knots and links [Ale28]. Reidemeister in 1927 and, independently, Alexander and Briggs in 1926, demonstrated that two knot diagrams of the same knot, up to an isotopy, can be related by a sequence of the three Reidemeister moves [Rei27, AB26], which will be precisely defined in Chapter 1. Thereafter, Thurston, Jones, Witten, and Kontsevich discovered and invented subtle connections between the classification of knots and links, via polynomial invariants, and partition functions of models in statistical mechanics and topological quantum field theory as well.

New developments on topological quantum field theory [Ati91] opened new possibilities, including models of loop quantum gravity and quantum information theory [Ash92, BB11, FKW02, KR19, Rov88, Wit89, Roc21, BO18]. One of the relationships between knot theory and quantum gravity resides on Chern-Simons theory and the derivation of link invariants, such as the Jones polynomial, from Yang-Mills theory [Bae94, BBG16, PR15]. Besides, knot invariants emerge naturally in gravitational physics, in the connection between knot theory and the loop representation of loop quantum gravity. Loop representations of quantum gravity on lattices were also proposed [LK06]. The Jones polynomial was expanded in terms of the cosmological constant, providing an infinite string of knot invariants that are solutions of the vacuum Hamiltonian constraint of quantum gravity in the loop representation [FGP96].

Interest in knot theory from the general mathematical community grew significantly after Jones discovered the Jones polynomial, in 1984 [Jon83, Jon85], yielding him to be awarded the 1990 Fields medal. This led to the derivation of other relevant knot polynomials, as the bracket, the HOMFLY, and the Kauffman polynomials. Thereafter, in 1988, Witten proposed a new framework for the Jones polynomial, utilizing existing ideas from mathematical-physics, such as Feynman path integrals, and introducing new notions such as topological quantum field theory [Wit89]. Witten also received the 1990 Fields medal, partly for this work. Witten's approach of the Jones polynomial yielded new invariants for the classification of 3-manifolds. Simultaneously, other approaches resulted in the Witten-Reshetikhin-Turaev invariants and various so-called quantum invariants, consisting of a mathematically more formal version of Witten's invariants [Tur94]. Latterly, fundamental particles in physics have been modelled as embroideries on weaves [Iwa92]. Nowadays, the most practical classification of knots and links is obtained via polynomial invariants.

In the 1980s, Conway discovered a procedure for unknotting knots, known as Conway notation [Con70]. In the early 1990s, knot invariants encoding the Jones polynomial and its generalisations, called the finite type invariants, were introduced by Vassiliev and Goussarov [Vas90], and also investigated in Refs. [BX93, BN95]. These invariants, previously described by classical topological tools, were shown by the 1994 Fields medalist Kontsevich to result from the integration of algebraic structures [Kon93]. Besides, Faddeev and Niemi derived knot-like structures as stable finite energy solitons where string-like structures appear, including physics of fundamental interactions, the early universe cosmology, and nematic liquid crystals [Fad96].

Figure 1: Knotting in DNA under an electron microscope [WDC85, Ada01, Sum95].

The discovery of polynomial invariants in such a subject as the structure of spacetime, so central to physics, and their relationship with vacuum expectation values in some field models [YG89], has made anyone wonder about it. Besides, the DNA molecule is well known to consist of two polynucleotide strands twisted around each other in a double helix, encoding life. It has several twists in it, as it coils around itself, consisting of a phenomenon called supercoiling [WDC85, Ada01, Sum95]. Polynomial invariants have central importance to study the knotting and packing/unpacking of the DNA molecule in Fig. 1.

In current language, the word *knot* is loosely used for proper knots, links, sometimes braids, and even more general weaving patterns. More precise notions are needed for mathematical discussions and will be given throughout this book. They correspond, however, to the popular notions related to those names.

The *overhand knot* and the *8-knot*, as can be found by experimenting with a string. They are different, in the sense that it is impossible to transform one into another without somehow tying or untying, that is, passing one of the ends of a strand through some loop. Some ways to tie the overhand knot include the thumb method, creating a loop and push the working end through the loop with your thumb, and also the overhand method, creating a bight, by twisting the hand over at the wrist and sticking your hand in the hole, pinch the working end with your fingers and pull through the loop. To forbid untying, one can either extend the ends to infinity, or simply connect them. We choose the latter, obtaining the closed versions, drawn symmetrically, as in Fig. 2. Besides, Figs. 3 and 4 illustrate, respectively, the crossing in a knot and the torus knots with a different number of crossing and links.

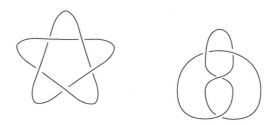

Figure 2: The cinquefoil and the 8-knot.

It is worth mentioning that the cinquefoil knot was named after herbs and shrubs of the rose family, presenting five-lobed leaves and five-petaled flowers [Wol18].

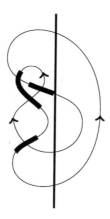

Figure 3: A weaving pattern of a knot.

The torus knots are an infinity family of prime knots which have particularly simple properties. The intimate relationship between braids and knots can be better shown in the case of torus knots. Given two coprime integers p and q, the torus knot $T_{p,q}$ can be constructed by wrapping a closed curve around the surface of a torus such that it encircles it p times along the meridian and q times along the longitude, defined respectively as the short and the long ways around the torus. A torus knot is trivial, equivalent to the unknot, if and only if either p or q is equal to 1 or -1. Torus knots are completely characterised p and q [Ban09]. They are invertible and chiral, e. g., they are not isotopic to their mirror image. The trefoil knot is equivalent to $T_{3,2}$. In Figs. 4 and 5, we illustrate some particular cases, respectively in two and three dimensions.

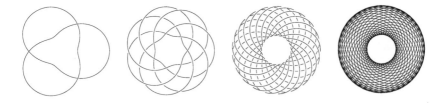

Figure 4: Torus knots, with different number of crossings and links. From left to right: the first plot displays the $T_{3,5}$ torus knot, whereas the second plot regards $T_{5,7}$. The third plot shows $T_{11,21}$ and the last plot illustrates the $T_{33,43}$ torus knot.

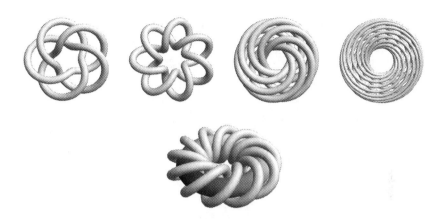

Figure 5: 3-dimensional torus knots, with different number of cross-
         ings and links. From left to right. (On the top): the first
         plot displays the $T_{3,5}$ torus knot, whereas the second plot
         regards $T_{2,7}$. The third plot shows $T_{7,6}$ and the last plot
         illustrates the $T_{17,7}$ torus knot; (at the bottom): the last
         figure below shows a torus knot sitting on a torus.

They change their names and, besides, become closed one-
dimensional compact spaces. Mathematicians assert that there
exists only one kind of such space, consisting of circles (or families of circles). Mathematically, a knot is defined as any subspace
of the Euclidean three-dimensional space $\mathbb{E}^3$ which is topologically equivalent to the circle (the circle is identified to the one-
dimensional S^1 sphere). This means that a knot is essentially a
circle, plunged in $\mathbb{E}^3$. In a more technical language, one asserts
that the knot is *homeomorphic* to the circle. A homeomorphism,
which implies a topological equivalence between two spaces, is
a continuous mapping, whose inverse mapping is also continuous. When such a mapping exists between two spaces A and B,

all the topological properties of A and B are the same. For instance, they have the same dimension. Thus, as spaces, all knots are topologically equivalent, consisting basically of circles. This may seem strange, as we are quite aware of differences between knots. Two knots, once closed, are equivalent, when it is possible to pass from one to the other by tying and untying, without ever cutting a strand. They are different when this procedure is not accomplishable. For example, the knot in Fig. 6 is trivial. It is equivalent to the circle itself, called the unknot. However, neither the trefoil nor the 8-knot can be unknotted, and they cannot be transformed into each other.

Figure 6: A trivial knot, and the unknot.

The bible on knots in practical life, Ref. [Ash93], lists 3,900 kinds of useful ones, navigating from the competence ranging from the sailor, through the knitter, the gardener, and the alpinist, to the fisherman and... the hangman as well, among others. One also cannot underestimate the importance of knots in medicine and nursery. Employing a needle and a thread, surgical knots are the only way to secure any suture. The so-called surgeon's knot is a slightly modified reef – or Hercules – knot, when one adds an extra twist when tying the first throw, resulting in a double overhand knot. More recently, a wonderful compendium, that lists 1,701,936 prime knots with up to 16 crossings, provides a complete and illustrative source of examples, in an intuitive presentation [HTW98].

In what do knots differ? The answer is that they are equivalent or different according to the way the circle is plunged in $\mathbb{E}^3$. A precise mathematical definition of the equivalence is embodied in the notion of *isotopy*, to be seen later. Links are submanifolds of $\mathbb{E}^3$, diffeomorphic to a disjoint union of circles, which are usually called components of the link. In some sense, links are collections of knots, that can be either interlaced or not. A knot is a link with just one component. Braids are quite well characterised algebraically, as they form groups. They are the best known of weaving patterns, and they allow furthermore a characterisation of links, and therefore knots, as their particular cases. A whole classification of weaving patterns, including carpets and spider webs, therefore results. In this book, we are concerned to reveal and provide a gentle introduction to knots, links, and braids. Braids have by themselves many applications in physics. It is thus natural to begin with them, delving into this fascinating structure.

This book consists of five chapters, each one ending with plenty of exercises for the reader who would like to weave a deeper knowledge and fix the concepts presented. Few figures throughout the text use Wolfram Research, Inc., Mathematica, Version 11.3.0.0, Champaign, IL (2018), and some other figures were depicted with the help of XYpic [Fis00]. In Chapter 1 the main fundamental concepts and prominent results on the elementary group theory are revisited, also to fix the notation for use throughout the book. Then knots are introduced, with the presentation of the braid group generators. Reidemeister moves are also studied, as well as the Burau and the direct product representations. A brief analysis of identical particles is accomplished, where the Yang-Baxter equation is presented in detail.

Chapter 2 is devoted to the introduction of types of knots and links, their relationship and features as well, as the writhe, twisting, and linking numbers. Topological structures, necessary to the understanding of the subsequent content in the book, are straightforwardly and intuitively introduced, without losing the formal aspects. It is carefully accomplished so that the reader can be endowed with some powerful tools that will be used in the chapters thereon. Topological spaces, homotopy, isotopy, and the fundamentals of homology, simplexes, and simplicial complexes are introduced, hence originating the Betti numbers and their correspondence to the Euler character. Some classical examples are presented, with important applications. The Reidemeister moves are shown to be generators of link isotopies. Also, Markov and Alexander's theorems are introduced and applied in this context.

The polynomial invariants are defined in Chapter 3, where prominent properties are presented in the context of the skein relations. Some examples of polynomial invariants associated with braids, knots, and links are introduced. The Conway formula, the Kauffman decomposition, and the Jones, Kauffman, HOMFLY, and Alexander polynomials are presented, as well as the Birman-Wenzl-Murakami one. Therefore, von Neumann algebras illustrate important relations among knots, links, and braids. The Kauffman diagrams are still explored in the context of the S matrix. Then the Manin quantum plane is briefly presented and discussed.

In Chapter 4 some models in statistical mechanics are presented and studied, which subsequently are going to be used in the description of quantum gases. The Ising model and the spontaneous symmetry breaking mechanism are presented. Besides, Weyl operators are studied in detail, as well as the Fourier transforms in $\mathbb{Z}_N$ and the Pontryagin duality. Their intimate relationship with the Heisenberg group is focused, on using the Schwinger basis. The Potts model and the star triangle are illustrated in the context of Cayley trees and Bethe lattices. Weyl operators are then shown to play a fundamental role in the Potts model. Finally, quantum gases are introduced by the braid group statistics.

Throughout this book the notation $\mathbb{N}, \mathbb{Z}, \mathbb{Q}, \mathbb{R}, \mathbb{C}$ will be used for denoting, respectively, the set of natural numbers, the rational numbers, the ring of integers, and the field of the real and complex numbers. The symbol $\mathbb{K}$ denotes an arbitrary field.

We acknowledge the support of the National Council for Scientific and Technological Development (CNPq), Coordination for the Improvement of Higher Education Personnel (CAPES), and The São Paulo Research Foundation (FAPESP), for academic foment over the years during which this work has been accomplished. The authors also thank Prof. Julio M. Hoff da Silva, for useful discussions and Dr. Anderson Alves da Silva, for helping the implementation of some figures throughout the text.

Chapter 1

Braids

1.1 General aspects

Braid groups concern real, usual braids. They count among the simplest examples of groups appearing in everyday life. One can easily build braids in practice, multiply and invert them. They are related to the study of general woven fabrics, which also includes knots and links. One of the standard mathematical references is Ref. [Bir75]. Fig. 1 displays a general example of a 4-braid.

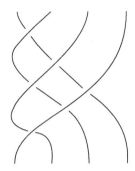 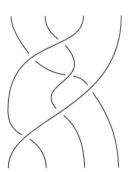

Figure 1: Examples of 4-braids.

Given two planes $\mathbb{E}^2 \subset \mathbb{R}^3$, with three chosen distinct points on each plane, it suffices to join these points in any way with strings, to come across with a braid. More precisely, let us consider two parallel planes $A, B \subset \mathbb{R}^3$, each one of them containing n distinct points in the sets $\{a_i\}_{i=1}^n$ and $\{b_j\}_{j=1}^n$, respectively. A n-strand braid is a collection of n curves $\{x_k\}_{k=1}^n$ such that

(1) each curve x_k has one endpoint, at the set $\{a_i\}$, and another endpoint, at the set $\{b_j\}$,

(2) all the x_i are pairwise disjoint,

(3) every plane parallel to A and B either intersects each of the x_i at one point or does not intersect at all.

Fig. 2 depicts some simple threads of 3-strands. Plot 2a shows a trivial 3-braid, with no interlacing of strands. Both the plots 2b and 2d illustrate the elementary steps of weaving, corresponding to two of the simplest nontrivial braids. By historical convention, strings are considered as going from the top to the bottom. Notice that, in the drawing, the plane $\mathbb{E}^2$ is represented by a

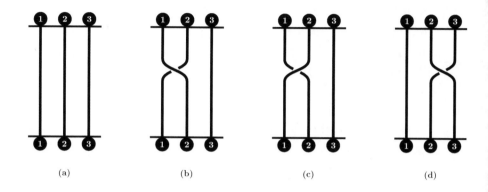

Figure 2: Some simple braids of 3-strands.

line, merely for the sake of simplicity. In the plot 2b, the line from the point labeled by 2 to the one labeled by 1 goes down behind that from 1 to 2. Just the opposite occurs in the plot 2c. Braids 2b and 2c are different, as they are thought to be drawn between two planes, so that an additional dimension, needed to make strings either overcrossing or undercrossing each other, is available.

The trivial braid 2a represents the neutral element, as the product between it and any other braid remains unaltered. Every braid has an inverse. The braids 2b and 2c are easily verified to be the inverse of each other. The very definition of the product, by concatenation, ensures associativity. The product of braids is clearly non-commutative. One can compare the braid 2b to 2d; and 2a to 2c. Indeed, any 3-strand braid can be obtained by successive multiplications of the elementary braids 2b, 2d, and their inverses. Such elementary braids are asserted to *generate* the third braid group, which is denoted by B_3. The procedure of building up braids as products of elementary braids can be used indefinitely. The braid group has consequently infinite order. Each braid can be interpreted as a mapping $\mathbb{E}^2 \to \mathbb{E}^2$, leading distinguished points into distinguished points, and the braid 2a as the identity mapping.

Braids can be multiplied by downward concatenation. Given two braids A and B, their product AB, denoted here by juxtaposition, is obtained by drawing B below A. Fig. 3 shows the product of the previous Fig. 2b by itself.

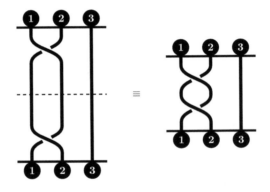

Figure 3: Product of the braid (2b) by itself.

Fig. 4 depicts the product of 2c and 2d, in Fig. 2.

This reasoning can be easily generalised to the nth braid group, B_n, also known as the Artin braid group, whose elements are braids with n-strands. It consists of the group whose elements are equivalence classes of n-braids, and whose group operation is the composition of braids. The best way to get used to the intuition regarding what is exposed here is to proceed modestly to real experiments with a few strings.

It is instructive to compare this point of view, with two copies

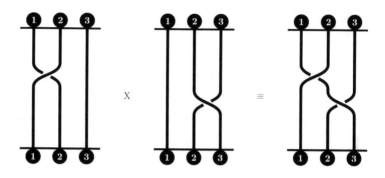

Figure 4: The product of the braids (2c) and (2d).

of the plane $\mathbb{E}^2$ embedded in the Euclidean host space, $\mathbb{E}^3$, to another one, purely restricted to the plane. Let us identify the two planes and consider the braid in Fig. 3. Each distinguished point is, ultimately, sent back into itself. Hence, it would appear, at first sight, that it corresponds to the identity, but that is not the case. In fact, the identity is represented by 2a, and the braid in Fig. 3 cannot be *unwoven*, reducing to it. Experiments show that it would be possible to disentangle it in the $\mathbb{E}^3$ space, however, it cannot be disentangled in $\mathbb{E}^2$. Since every braid is a composition of elementary braids, that would mean that any braid on $\mathbb{E}^3$ can be unbraided, as witnessed by millennia of practice with hair braids. Hair braids in $\mathbb{E}^2$ can be simulated by somehow gluing together their extremities, thereby eliminating one degree of freedom.

Since braids can be unwoven in $\mathbb{E}^3$, the braid group reduces to the symmetric group, and quantum and statistical mechanics in $\mathbb{E}^3$ remain what they are usually. Differences could however appear in the two-dimensional case. Anyhow, from the intrinsic

$\mathbb{E}^2$ point of view, what seems to be the identity exhibits infinite possibilities, as it could also be obtained by the composition of any number of the braids in Fig. 3. Let us consider again Fig. 3, but now as it would be seen when projected on $\mathbb{E}^2$. Entwining in $\mathbb{E}^3$ reduces to oriented exchanges in $\mathbb{E}^2$. The braid 2b represents a counterclockwise exchange of points 1 and 2 (Fig. 2a). By the braid in Fig. 2b, both points come back to their original position. Strings are nevertheless impenetrable and cannot cross each other. The strings impenetrability can be simulated by representing each strand by a hole in $\mathbb{E}^2$, a point that is forbidden to any other hole. Repeated multiplication by the braid in Fig. 2b leads to paths turning $2, 3, \ldots, n$ times. Each particle sees the others as forbidden points, like holes. All this strongly suggests a relation to the fundamental group of the punctured $\mathbb{E}^2$. It is indeed through this relation that mathematicians approach braid groups, as seen below.

The multiplicity of the identity braid is merely a particular case and would be simply twice the transposition of points 1 and 2. The nth braid group B_n is an enlargement of the group of permutations, S_n. Mathematicians have several definitions for B_n, the previous configuration space definition allowing, as asserted, the generalisation of braid groups on any manifold M. The double exchange of two particles is usually supposed to lead to identity permutation. However, as already discussed, this may be different in two-dimensional space. Once defined, a few basic properties regarding braids need then to be exposed, such as what it means for two braids to be equivalent and that braids form a group under the operation of composition, called the Artin braid group. Such a prominent group can be defined using simple generators and relations, presenting relevant algebraic properties. In what follows, we also aim to pave the way to construct the correspondence between braids and knots and show how the Jones polynomial for knots can be derived from a representation of the corresponding braid group.

1.2 Permutation groups

Let $A = \{a_1, a_2, \ldots, a_n\}$ be a finite set of n elements, where the a_j are called *letters*. A *permutation* of A is a bijection $A \to A$ [Fra74]. The usual notation for a fixed permutation, wherein each a_j is led into some a_{p_j}, reads

$$\begin{pmatrix} a_1 & a_2 & \ldots & a_j & \ldots & a_{n-1} & a_n \\ a_{p_1} & a_{p_2} & \ldots & a_{p_j} & \ldots & a_{p_{n-1}} & a_{p_n} \end{pmatrix}. \tag{1.1}$$

The set of all permutations of the set A constitutes a group, under the operation of composition, i. e., the product, called the nth *symmetric group*, denoted by S_n. Taking the particular case when $n = 4$, an example of a product in S_4 is given by

$$\begin{pmatrix} a & b & c & d \\ b & d & c & a \end{pmatrix} \begin{pmatrix} a & b & c & d \\ a & c & b & d \end{pmatrix} = \begin{pmatrix} a & b & c & d \\ c & d & b & a \end{pmatrix}. \tag{1.2}$$

Drawings are prominently conductive in understanding the notation. Fig. 5 shows the product (1.2), where the composition is similar for braid groups. Notice solely that the right action is used to comply with the downward convention for braids.

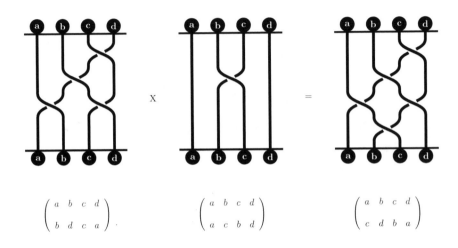

Figure 5: An example of permutation product in S_4.

There is a fundamental difference between the braiding and the permutation group representations. In S_n, strands do not differ between undercrossing and overcrossing. The order, namely the number of elements of S_n, is given by $|S_n| = n!$. A permutation having the special form

$$\begin{pmatrix} a_1 & a_2 & \ldots & a_{r-1} & a_r \\ a_2 & a_3 & \ldots & a_r & a_1 \end{pmatrix} \tag{1.3}$$

is a *cycle* of length r, usually denoted by $(a_1, a_2, \ldots, a_r)$. A product of two cycles is not necessarily a cycle. The product of disjoint cycles (cycles with no letter in common) is commutative.

A very important fact is that every permutation can be written as a product of disjoint cycles. Eq. (1.2) can be written as

$$(a, b, d)(c)(a)(b, c)(d) = (a, c, b, d). \tag{1.4}$$

It is convenient to attribute a variable t_r to a cycle of length r and indicate the cycle structure of a permutation by the monomial $t_1^{n_1} t_2^{n_2} t_3^{n_3} \ldots t_r^{n_r}$, meaning that there are n_1 1-cycles, n_2 2-cycles, $\ldots$, n_r r-cycles [Ham62]. In the example (1.4), the monomial is $t_1^3 t_2 t_3$.

Permutations, P, of the same cycle type, namely, with the same set $\{n_j\}$ of exponents, can be led to each other under the adjoint action, SPS^{-1}, for any element $S \in S_n$. This means that permutations are representatives of conjugate classes. Hence, one monomial is attributed to all permutations of a fixed class. In this sense, such monomials are invariants of the group S_n. It should be noticed that each class corresponds also to a representation of the group, so that there is a monomial for each representation. A polynomial corresponds therefore to a sum over representations.

A cycle of length 2 is a *transposition*, represented by the cycle (a, b). Every cycle is a composition of transpositions, so that every permutation can be written ultimately as a product of transpositions. Transpositions generate the symmetric group. A basis of a group G is a set of elements $\{b_k\}_{i=1}^n \subset G$ such that any $g \in G$ can be obtained as a product of elements in $\{b_k\}$. The elements of $\{b_k\}$ are then called generators of G. Permutations formed by the product of an odd [even] number of transpositions are called *odd [even] permutations*. Using the notation

$$
\begin{pmatrix}
1 & 2 & \ldots & n-1 & n \\
p_1 & p_2 & \ldots & p_{n-1} & p_n
\end{pmatrix},
\tag{1.5}
$$

one can use as a basis of S_n the set $\{s_j\}_{j=1}^n$ of elementary transpositions. The element s_i exchanges only the ith and the $(i+1)^{\text{th}}$ entry,

$$
s_i =
\begin{pmatrix}
1 & 2 & \ldots & i & i+1 & \ldots & n-1 & n \\
1 & 2 & \ldots & i+1 & i & \ldots & n-1 & n
\end{pmatrix},
\tag{1.6}
$$

obeying the relations, for all $\{i, j\} \subset \{1, \ldots, n\}$,

$$s_i s_{i+1} s_i = s_{i+1} s_i s_{i+1}, \tag{1.7}$$

$$s_i s_j = s_j s_i, \qquad \text{for } |i - j| \geq 2, \tag{1.8}$$

the first of which is pictured in Fig. 6, for $n = 4$. Notice the

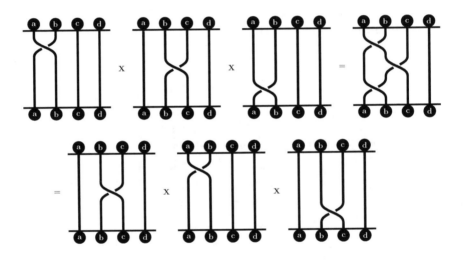

Figure 6: Pictorial presentation of the relation $s_1 s_2 s_1 = s_2 s_1 s_2$.

behavior of the third strand in Fig. 6. It has been exchanged twice with the second one and, consequently, it appears as a simple, unmoved strand. This comes from the property

$$s_i s_i = \text{id}, \tag{1.9}$$

typical of transpositions. Indeed, complying with straightforward accessible experiments, braid groups generators satisfy relations (1.7, 1.8), but there is no analog to (1.9) for them. This property means precisely that a pure permutation does not descry between a strand going over or under the other. Each elementary permutation s_i is identical to its inverse.

It is sometimes interesting and useful to deal with matrix representations of S_n. The most usual basis consists of matrices that, when applied to column vectors, simply exchange consecutive entries. For $n = 4$, for instance, one can represent

$$s_1 = \begin{pmatrix} 0&1&0&0 \\ 1&0&0&0 \\ 0&0&1&0 \\ 0&0&0&1 \end{pmatrix}, \quad s_2 = \begin{pmatrix} 1&0&0&0 \\ 0&0&1&0 \\ 0&1&0&0 \\ 0&0&0&1 \end{pmatrix}, \quad s_3 = \begin{pmatrix} 1&0&0&0 \\ 0&1&0&0 \\ 0&0&0&1 \\ 0&0&1&0 \end{pmatrix}. \quad (1.10)$$

General group elements are represented by the corresponding matrices.

Another basis is more adequate for using in knot theory. The generators for S_n are given in terms of $(n+1) \times (n+1)$ matrices. In S_4, for example, they read

$$s_1 = \begin{pmatrix} 1&0&0&0&0 \\ 0&0&1&0&0 \\ 0&1&0&0&0 \\ 0&0&0&1&0 \\ 0&0&0&0&1 \end{pmatrix}, \quad s_2 = \begin{pmatrix} 1&0&0&0&0 \\ 0&1&0&0&0 \\ 0&0&0&1&0 \\ 0&0&1&0&0 \\ 0&0&0&0&1 \end{pmatrix}, \quad s_3 = \begin{pmatrix} 1&0&0&0&0 \\ 0&1&0&0&0 \\ 0&0&1&0&0 \\ 0&0&0&0&1 \\ 0&0&0&1&0 \end{pmatrix}. \quad (1.11)$$

These are isomorphic to the matrices appearing in (1.10). Other bases for representations can be obtained from that one by similarity transformations, which do not change the cycle character of any product. The monomial $t_1^{n_1} t_2^{n_2} t_3^{n_3} \ldots t_r^{n_r}$, corresponding to an element of the symmetric group, is invariant under similarities and is a cognizable example of polynomial invariant. A special basis, which will be useful later on, is obtained from the matrices (1.11) by similarity, induced by a left-lower triangular matrix, denoted here by C, whose non-vanishing elements are equal to 1. For $n = 4$, it reads

$$C = \begin{pmatrix} 1 & 0 & 0 & 0 & 0 \\ 1 & 1 & 0 & 0 & 0 \\ 1 & 1 & 1 & 0 & 0 \\ 1 & 1 & 1 & 1 & 0 \\ 1 & 1 & 1 & 1 & 1 \end{pmatrix}, \tag{1.12}$$

and the new basis is formed by conjugation,

$$s_1' = Cs_1C^{-1} = \begin{pmatrix} 1 & 0 & 0 & 0 & 0 \\ 1 & -1 & 1 & 0 & 0 \\ 0 & 0 & 1 & 0 & 0 \\ 0 & 0 & 0 & 1 & 0 \\ 0 & 0 & 0 & 0 & 1 \end{pmatrix}, \tag{1.13}$$

$$s_2' = Cs_2C^{-1} = \begin{pmatrix} 1 & 0 & 0 & 0 & 0 \\ 0 & 1 & 0 & 0 & 0 \\ 0 & 1 & -1 & 1 & 0 \\ 0 & 0 & 1 & 0 & 0 \\ 0 & 0 & 0 & 0 & 1 \end{pmatrix}, \tag{1.14}$$

$$s_3' = Cs_3C^{-1} = \begin{pmatrix} 1 & 0 & 0 & 0 & 0 \\ 0 & 1 & 0 & 0 & 0 \\ 0 & 0 & 1 & 0 & 0 \\ 0 & 0 & 1 & -1 & 1 \\ 0 & 0 & 0 & 0 & 1 \end{pmatrix}. \tag{1.15}$$

It corresponds to a reducible representation, as the last row and column of s_1' and s_2', and the first row and column of s_3', can be excluded with no change to the algebra respected by the generators. The cases of general n can be similarly obtained. From this basis, the Burau representation of the braid group can be obtained by a certain deformation.

1.3 Identical particles

Let us briefly revisit in few words how the symmetric group appears in elementary quantum mechanics of a system of n identical particles. All their coordinates, spins, and other quantities are denoted by the collective variable x. Exchanging particles is provided by the action of elements in S_n on x. General permutations are products of elementary transpositions of two particles, each one given by $s_i x$, for some $i \in \{1, \ldots, n\}$. For instance, if $n = 2$ and $x = (r_1, r_2)$, the unique possible exchange reads $(r_2,$

$r_1) = s_1[(r_1, r_2)]$. The states are given by rays in the Hilbert space H of wave functions. More precisely, states are representatives of the equivalence class determined by the equivalence relation between wave functions

$$\psi_1(x) \sim \psi_2(x) \quad \text{if} \quad \psi_2(x) = \lambda\psi_1(x), \qquad \lambda \in \mathbb{C}^*. \quad (1.16)$$

Equivalence classes for the relation $\sim$ are called rays or projective rays. In quantum mechanics, the projective Hilbert space $P(H)$ of a complex Hilbert space H takes $\lambda = e^{i\eta}$, where η is an arbitrary phase. The physical meaning of the projective Hilbert space is that, in quantum theory, the wave functions ψ and $\lambda\psi$ do represent the same physical state, for any $\lambda \neq 0$. It is conventional to choose a wave function ψ from the ray so that it has a unit norm, $\langle\psi|\psi\rangle = 1$, in which case it is called a normalised wave function. The unit norm constraint does not completely determine ψ within the ray, since ψ can be multiplied by any unitary λ, corresponding to the U(1) group action, preserving the normalisation. No measurement can recover the phase of a ray, and therefore the phase is not an observable, in quantum mechanics.

A group symmetry in quantum mechanics yields wave functions to respond to transformations according to a unitary (or anti-unitary) representation of the group. Therefore, under exchanges, the wave function changes through an unitary representation $U(S_n)$ of S_n in the Hilbert space. A permutation P is represented by an operator $U(P)$. The set $\{U(s_i)\}_{i=1}^n$ forms a basis for such a representation and respects conditions corresponding to (1.7) – (1.8). There is, however, an additional condition: as ψ is supposed to have values in a one-dimensional complex space, this representation must be a one-dimensional unitary representation, and consequently only phase factors will appear. In fact, for each $j \in \{1, \ldots, n\}$, one has

$$\psi(s_j x) = U(s_j)\psi(x) = e^{i\varphi_j}\psi(x), \quad (1.17)$$

for same phase φ_j.

It is immediate to see that Eq. (1.7) imposes the equality of all the phases, so that $U(s_j)\psi(x) = e^{i\varphi}\psi(x)$, with the same phase φ for all s_j. Analogously, Eq. (1.8) yields

$$\begin{aligned} U^2(s_j)\psi(x) &= U(s_j^2)\psi(x) = e^{i2\varphi}\psi(x) \\ &= \psi(x), \end{aligned} \tag{1.18}$$

implying that $e^{i\varphi} = \pm 1$. There are only two one-dimensional representations: the totally symmetric one, related to bosons and with $U(P) = +1$ for every permutation P; and the totally antisymmetric one, in which fermions find their place, with $U(P) = +1$, when P is even, and $U(P) = -1$, when P is odd. If x denotes only positions of the particles on a manifold M, the configuration space for the system of n particles corresponds to the product manifold $M \times M \times \cdots \times M = M^n$. For indistinguishable particles, it would consist of the quotient of that by the symmetric group, M^n/S_n. This is a multiply-connected space. In the usual $M = \mathbb{E}^3$ case, its fundamental group is $\pi_1(\mathbb{E}^{3n}/S_n) = S_n$ and $\mathbb{E}^{3n}$ is its universal covering.

Notice that the symmetric group S_n is specified altogether (up to isomorphisms) by asserting that it has $(n-1)$ generators satisfying relations (1.7 – 1.9). When a group is introduced in this way, by symbols (letters) representing its generators and some equations they must satisfy, this group is said to be given by a *presentation*. This has some analogy with the introduction of surfaces in a Euclidean space, when coordinates are given and then subject to defining relations. A parenthesis on this subject will be in a good place here.

1.4 Free groups

Let us start with free groups, or word groups [Fra74, CF63]. Going back to the set $A = \{a_1, a_2, \ldots, a_n\}$, we have called letters the elements a_j and will now call A itself an *alphabet*. An element with p consecutive times the letter a_j is written as a_j^p and it is called a *syllable*. A finite string of syllables is a *word*. The *empty word* 1 has no syllables. On a given word there are two types of transformations, called *elementary contractions*. They correspond to the usual manipulations of exponents. By a contraction of the first type, symbols like $a_j^p a_j^q$ are replaced by a_j^{p+q}; by a contraction of the second type, a symbol like a_j^0 is replaced by the empty word 1, or simply dropped from the word. With these contractions, each word can be reduced to its simplest expression, a *reduced word*. The set $F[A]$ of all the reducible words of the alphabet A can be made into a group. The product $u \cdot v$ of two words u and v is just the reduced form of the juxtaposition uv. It is possible to show that this operation is associative and attributes an inverse to every reduced word. The resulting group $F[A]$ is the *free group generated by A*.

A general group G is called a free group if it has a set $A = \{a_1, a_2, \ldots, a_n\}$ of generators such that G is isomorphic to $F[A]$. In this case, the a_j are the *free generators* of G. The number of letters is the *rank* of G and may be eventually infinite. The importance about free groups comes from the following theorem:

> ☞ *Every group* G *is a homomorphic image of some free group* $F[A]$.

This means that a mapping $f : F[A] \to G$ exists, preserving the group operation. A homomorphism, in general, loses something: many elements in $F[A]$ may be taken into the same element of G. Therefore, $F[A]$ can be oversised. To obtain an isomorphism, something else must be implemented, extracting the excess through relations between elements of $F[A]$. Another version of the previous theorem is:

> ☞ *Every group G is isomorphic to some quotient group of a free group.*

A subset N of G is called a subgroup of G if N also forms a group under the binary operation that defines G. The subgroup N is called a normal subgroup if it is invariant under conjugation by elements of G. Namely, a subgroup N is normal in G if and only if $gng^{-1} \in N$, for all $g \in G$ and $n \in N$. The usual notation for this relation is $N \lhd G$. Normal subgroups are important because only they can be used to construct quotient groups of the given group. Normal subgroups of G are precisely the kernels of group homomorphisms with domain G, which means that they can be used to classify those homomorphisms. Let us consider a subset $\{r_j\} \subset F[A]$. One builds the minimal normal subgroup R with the r_j as generators. The quotient $F[A]/R$ is a subgroup, corresponding to putting $r_j = 1$. An isomorphism of G onto $F[A]/R$ is a presentation of G. The set A is the set of generators and each r_j is a *relator*. Each $r \in R$ is a *consequence* of $\{r_j\}$. Each equation $r_j = 1$ is a *relation*.

Groups are introduced by defining generators and relations among them. Free groups have for discrete groups a similar role to that of coordinate systems for surfaces: these are given, in a larger space, by the coordinates and related equations.

There are two ways in which the symmetric group can be related to free groups:

1) one can think of the generators s_i as letters. Any element of S_n corresponds to a word like $s_1^{q_1} s_2^{q_2} s_3^{q_3} s_4^{q_4} \ldots s_1^{p_1} s_2^{p_2} \ldots$, with Eqs. (1.7 − 1.9) consisting of relations;

2) the letters are elements of the set $A = \{a_1, a_2, \ldots, a_n\}$ and their permutations are seen as automorphisms on the group they constitute.

This may not be the simplest way to introduce S_n, but many groups are only defined through a presentation. Anyhow, this is frequently the better way to introduce discrete groups. It is the case of braid groups. As a curiosity, the complete classification of all two-dimensional topological manifolds, which has been performed in terms of series of discrete groups, can be compactly encoded into words [DNF79].

1.5 Artin classical braids

There are several definitions for the braid group B_n, some of them of great physical interest. Given the interval $I = [0, 1] \subset \mathbb{R}$ and $j \in \{1, 2, \ldots, n\}$, a braid is a family of distinct, non-intersecting, curves $\{\gamma_1, \gamma_2, \ldots, \gamma_n\}$, where the $\gamma_j : I \to \mathbb{E}^2 \times I$ are defined in such a way that with

$$\gamma_j(0) = (P_j, 0), \tag{1.19}$$
$$\gamma_j(1) = (P_{\sigma(j)}, 1), \tag{1.20}$$

where $\{P_1, \ldots, P_n\}$ are n distinct points in the plane $\mathbb{E}^2$ (one usually assumes $P_j = (j, 0) \in \mathbb{E}^2$), and σ is an index permutation. A braid is called a *tame* braid when its curves have continuous first-order derivatives, namely, when they are of class C^1. Otherwise, the braid is asserted to be a *wild* braid. In Fig. 7, the left plot shows again the elementary braid 2b, whereas the plot on the right represents the identity braid.

Mathematicians call braids, when introduced in this way, by Artin's braids [Art47, Boh47] or geometrical braids. The braid group B_n consists now of compositions of path meshes. Those braids corresponding to the identity permutation, as that of Figs. 2a and 3, are called *colored* or *pure* braids.

Let us be more precise about this correspondence. There is a surjective homomorphism of the braid group into the symmetric

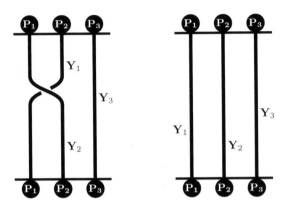

Figure 7: Geometrical braids, seen as curves on $\mathbb{E}^2 \times I$.

group,

$$h : B_n \to S_n, \tag{1.21}$$

sending a braid of n strands to its associated permutation. The centre, $\ker h$, of this homomorphism, consists of the subgroup of braids of B_n corresponding to the trivial permutation and is therefore composed of colored (or pure) braids. This homomorphism erases the differences coming from strings that either overcross or undercross each other.

The n-strand braid group consists of equivalence classes of braids, with the composition operation connecting the bottom of the first braid to the top of the second, and subsequently rescales the result to preserve unit length. The trivial braid has n parallel strands, and the inverse of a braid can be thought of as being its mirror image. One can produce a knot or a link by connecting the corresponding top and bottom strands by paths in $\mathbb{E}^3$, yielding a closed braid. For the n-strand group, one can use a basis

$\{\sigma_j\}_{j=1}^{n-1}$ of generators which are led by this homomorphism into elementary transpositions, that is, such that $h(\sigma_j) = s_j$. They obey the relations

$$\sigma_j\sigma_{j+1}\sigma_j = \sigma_{j+1}\sigma_j\sigma_{j+1}, \qquad \text{for} \quad 1 \le i \le n-2, \qquad (1.22)$$

$$\sigma_i\sigma_j = \sigma_j\sigma_i, \qquad \text{for} \quad |i-j| > 1, \qquad (1.23)$$

which can be alternatively used as a definition of the group B_n. Therefore, the braid group is so introduced by a presentation. This is the most convenient definition, from a computational point of view. Relations (1.23) can be easily seen to be impelled from the drawings of Sect. 1.2. In fact, any generator $\sigma_i \in B_n$ corresponds to exchanging the ith-strand with the $(i+1)^{\text{th}}$-strand. The line starting at i crosses above the another one, whereas the remaining lines go just straight.

The group B_1 consists of all braids constructed upon one strand, then consisting of the trivial group itself. Besides, B_2 is constituted by twists of two strands. One can associate a positive sign with a twist in one direction, whereas with a twist in the opposite direction it is associated a negative sign. Therefore, B_2 is isomorphic to the group of integers, endowed with the addition operation.

The braid group B_3 is given by the presentation

$$B_3 = \langle s_1, s_2 \,|\, s_1s_2s_1 = s_1s_1s_2 \rangle. \qquad (1.24)$$

Fig. 8 illustrates the generators of the braid group B_3.

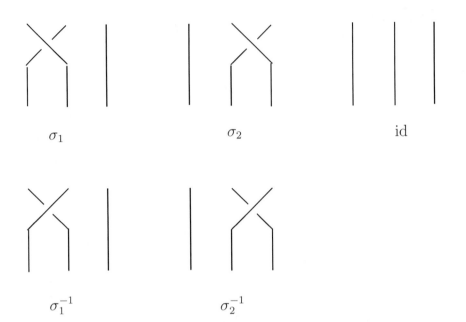

$$\sigma_1 \qquad\qquad \sigma_2 \qquad\qquad \mathrm{id}$$

$$\sigma_1^{-1} \qquad\qquad \sigma_2^{-1}$$

Figure 8: Generators $\sigma_1, \sigma_2 \in B_3$, the identity, and their inverses.

The drawings obtained by composing these elementary braids verify automatically the braid relations. Any braid can thus be written as a product of powers of the generators. Hairdos on a head with n hairs are thus encoded in B_n and comprise n-strands. The group B_n has the following presentation, following Eqs. (1.22, 1.23):

$$B_n = \langle \sigma_1, \ldots, \sigma_{n-1} \mid \sigma_i\sigma_{i+1}\sigma_i = \sigma_{i+1}\sigma_i\sigma_{i+1}, \sigma_i\sigma_j = \sigma_j\sigma_i \rangle, \quad (1.25)$$

where in the first group of relations one must observe that the indexes are constrained to $1 \leq i \leq n-2$, whereas the second group of relations is constrained to $|i-j| \geq 2$. The cubic relations, known as the braid relations, play an important role in the theory of

Yang-Baxter equations. In general, the braid group B_n generators can be represented by the braiding pattern in Fig. 9.

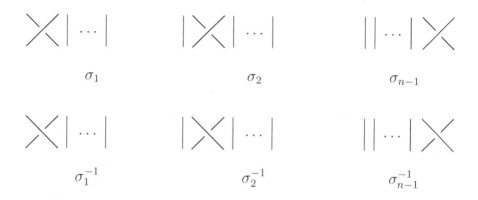

Figure 9: The braid group B_n generators.

One calls any sequence of elements $\sigma_i^{\pm 1}$ a braid word. If a braid word contains only elements of type σ_i and no elements of type σ_i^{-1}, then it will be called a positive braid word. If only elements of type σ_i^{-1} exist on a braid word, with no elements of type σ_i, then it is called a negative braid word.

Returning to torus knots, one can realize that the $T_{p,q}$ torus knot is equivalent to the $T_{q,p}$ torus knot, showing by moving the strands on the surface of the torus. Also, the $T_{p,-q}$ torus knot is the mirror image of the $T_{p,q}$ and the $T_{-p,-q}$ torus knot is equivalent to the $T_{p,q}$ torus knot but with reversed orientation [LR97, Liv93, Kaw96]. Any $T_{p,q}$ torus knot can be constructed from a closed braid with p strands, with braid word

$$(\sigma_1 \sigma_2 \cdots \sigma_{p-1})^q. \tag{1.26}$$

Fig. 10 illustrates some particular cases.

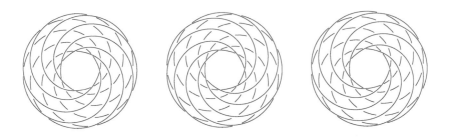

Figure 10: Torus knots, with different number of crossings and links. From left to right: the first plot displays the $T_{9,7}$ torus knot, whereas the second plot regards $T_{9,-7}$. The third plot shows $T_{-9,-7}$.

Ref. [MT92] used generalised Burau matrices with entries in the ring of the braid group to iteratively produce new braid representations out of any given one. Besides, Ref. [CT92] linearised the Artin representation of the braid group given by automorphisms of a free group providing a linear faithful representation of the braid group. This result is generalised to obtain linear representations for the pure braid group. Applications to some areas of two-dimensional physics are discussed.

1.6 Braid group generators and braid statistics

It is worth to notice that, unlike the elementary exchanges of the symmetric group, the square of an elementary braid is not the identity. Going back to quantum mechanics, a basis for a representation of a braid group is given by operators $U(\sigma_j)$ acting on wave functions, according to $U(\sigma_j)\psi(x) = e^{i\varphi}\psi(x)$. However, now there is no constraint enforcing $U(\sigma_j^2) = 1$, so that

$$U^2(\sigma_j)\psi(x) \;=\; U(\sigma_j^2)\psi(x) = e^{i2\varphi}\psi(x). \qquad (1.27)$$

Hence, using finite induction, one can straightforwardly prove that

$$U(\sigma_j^k)\psi(x) = e^{ik\varphi}\psi(x), \qquad\qquad \forall\, k \in \mathbb{N}. \qquad (1.28)$$

The representation is now, like the group, infinite. It is from the condition $U(\sigma_j^2) = 1$ that the possibilities of values for the phase, for usual n-particle wave functions, are reduced to two. In fact, since twice the same permutation leads to the same state, $U(\sigma_j^2)\psi(x) = \psi(x)$, it yields $e^{i\varphi} = \pm 1$. The two signs correspond to wave functions which are either symmetric or antisymmetric under exchange of particles, when one considers, respectively, bosons or fermions. When statistics is governed by the braid group, as it is the case of two-dimensional configuration spaces of impenetrable particles, the phase $e^{i\varphi}$ remains arbitrary and there is a different statistics for each value of φ. Such statistics is called braid statistics.

The absence of the involutory relation (1.9) has, as already asserted, deep consequences. Unlike the elementary exchanges of the symmetric group, the square of an elementary braid is not the identity. In many important applications, however, σ_j^2 differs from the identity in a well-defined way. In the simplest case, σ_j^2 can be expressed in terms of the identity and σ_j, which means that it satisfies a second-order equation like $(\sigma_j - x\,\mathbf{1})(\sigma_j - y\,\mathbf{1}) = 0$, where x and y are numbers and $\mathbf{1} \in B_n$ denotes the identity mapping. In this case, the generators σ_j belong to a subalgebra of the braid group algebra, called Hecke algebra. This is the origin of the so-called skein relations, which are helpful in the calculation of polynomial invariants in knot theory.

Before proceeding, the definition of the loop is essential in what follows. A loop in a topological space X is a continuous function $f : I = [0,1] \to X$ such that $f(0) = f(1)$. In other words, it is a path whose initial point is equal to its terminal point [Ada78]. A loop can also be seen as a continuous mapping $f : S^1 \to X$, since S^1 can be thought of as a quotient of I under the antipodes identification of 0 with 1. The set of all loops in X is the loop space of X.

Up to now, braids living essentially in $\mathbb{E}^3$ have been studied by their diagrams and their projections on the plane $\mathbb{E}^2$. This is more important for knots, whose visualisation is more difficult. That is why a careful treatment of such projections is necessary. Some equivalences, taken for granted, can be punctiliously codified so that the equivalence between two drawings is obtained thoroughly. Such steps are called *Reidemeister moves* and are shown in Fig. 11, for the $n = 3$ case. A Reidemeister move is any of three local moves on a link diagram. Each move operates on a small region of the diagram and is one of three types.

A Reidemeister type-I move (RI) corresponds to put or take out a kink, twisting and untwisting in either direction. A Reidemeister type-II move (RII) consists of sliding a strand over [under] to create [remove] two crossings. It is equivalent to move one loop completely over another one. They simply straighten the strands. RII is indeed the relation of an elementary braid to its inverse. A Reidemeister type-III move (RIII) corresponds to using (1.22), related to sliding a strand across some crossing, or moving a string completely over or under a crossing. Later it will be clear that RIII, ultimately, is the essential content of the Yang-Baxter equation. In accomplishing any Reidemeister move, no other part of the diagram is involved, and a planar isotopy may distort the picture. The numbering that labels the types of moves corresponds to how many strands are involved. For instance, an RII operates on two strands of the diagram, whereas an RIII operates on three strands. Among all Reidemeister moves, RI is the only move that alters the writhe of the diagram. One assumes with each move that the diagram is only locally modified, leaving the rest of the diagram unchanged.

All this may seem trivial in the straightforward examples already presented, but the strict observance of a step-by-step procedure is essential to show the equivalence of intricate weaving patterns. It turns out that rules for only one, two, and three strands suffice to establish an isotopic relationship. The *Reidemeister theorem* states then that two links are isotopic-equivalent if and only if their diagrams can be obtained from each other by some finite series of RI, RII, and RIII moves represented in Fig. 11. In other words, two knots or link diagrams are topologically equivalent if and only if their projections may be deformed into each other by a sequence of RI, RII, and RIII moves and planar ambient isotopies. The main idea of the proof is that piecewise-linearly, one can reduce to the consideration of a few minimal moves on knot diagrams, which one shows can be obtained as a finite sequence of Reidemeister moves [Rei27, AB26]. A relevant context wherein the Reidemeister moves play a prominent role is in defining knot invariants. A property of a knot diagram that does not change, under Reidemeister moves, defines an invariant. As none of the Reidemeister moves alters the number of components of a link diagram, the number of components is, therefore, an isotopy invariant. In particular, one can conclude that the Hopf link and the Borromean rings are not isotopic.

Although in principle one can always find a sequence of Reidemeister moves between two equivalent link diagrams, there is no straightforward way of guessing which moves one must implement. A more difficult task consists of distinguishing two different knots using just Reidemeister moves. Knot invariants comprise a more direct way to answer these important questions. A knot invariant is any function of knots that depends only on their equivalence classes. Many important invariants can be defined in this way, including the celebrated Jones polynomial.

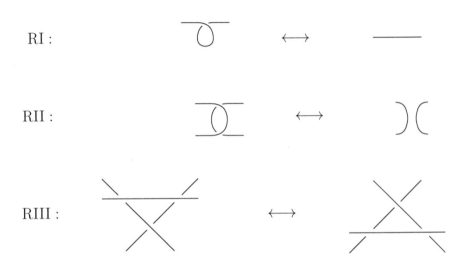

RI :

RII :

RIII :

Figure 11: Reidemeister moves.

Notice that RI and RII are simplifying steps, in the sense that they reduce the number of crossings in the diagram. We will later relate knots and links to braids, upon which Reidemeister moves are also applied to. Links are going to be characterised by invariant polynomials, and R-moves (Reidemeister moves) appear as the simplest way to demonstrate the isotopic invariance of a polynomial. Fig. 12 illustrates examples of such moves in the context of braids.

Figure 12: Equivalence between braids under RIII moves.

Besides, the two braids in Fig. 13 can be led to each other by the composition of the inverse of the first Reidemeister move RI. Subsequently, in the opposite side of the curve with respect to the another curve, after RI itself.

Figure 13: Equivalence between braids under RI moves.

1.7 Formal approach

A more formal approach views braid groups as fundamental groups of certain spaces. Curiously enough, this is the standpoint of allowing a closer connection to mathematics. Mathematicians prefer it, as allows generalisations to braid groups on any manifold M.

Consider n distinct particles on the Euclidean plane $M = \mathbb{E}^2$. Their configuration space $M^n = \mathbb{E}^{2n}$ reads

$$M^n = \{(x_1, x_2, \ldots, x_n), \ x_i \in M, \ \forall i \in \{1, \ldots, n\}\}, \quad (1.29)$$

the nth Cartesian product of manifold M by itself. Suppose that the particles are impenetrable, so that two of them cannot occupy the same position in $\mathbb{E}^2$. To take this into account, define the set

$$D_n = \{(x_1, \ldots, x_n) \mid x_i = x_j, \ \text{for some } \{i, j\} \subset \{1, \ldots, n\}\} \quad (1.30)$$

and consider its complement in M^n,

$$F_n M = M^n \backslash D_n. \quad (1.31)$$

It can be interpreted as M with n distinguished points, but with coincidences forbidden. Such would be the case of a gas of *impenetrable* particles. The fundamental group of this configuration space, $P_n = \pi_1[F_n M]$, is the *pure braid group*. By the way, the name colored braid comes from the fact that there exists a way of attributing a color to each strand which is preserved by the P_n group product.

Consider now identical and indistinguishable particles. The configuration space is still reduced: two points x and x' are equivalent if $(x_1, x_2, \ldots, x_n)$ and $(x'_1, x'_2, \ldots, x'_n)$ differ only by a permutation, a transformation belonging to the symmetric group S_n. Let $B_n M$ be the space obtained by identification of all equivalent points, the quotient by S_n,

$$B_n M = [F_n M]/S_n. \quad (1.32)$$

The fundamental group of this configuration space, $\pi_1(B_n M)$, is the *full braid group*, or simply *braid group*.

Artin's braid group is the full braid group for $M = \mathbb{E}^2$. It is now immediate to see that a colored geometrical braid is a representative of a class of loops on $F_n M$, that is, an element of $\pi_1[F_n M]$. A general geometrical braid is a representative of a class of loops on $B_n M$, an element of $\pi_1(B_n M)$. The name *configuration space* is used by mathematicians but physicists will recognise it immediately. The space $F_n M$ is just the configuration space for a gas of n impenetrable particles, and $B_n M$ is the configuration space for a gas of n impenetrable and identical particles. Therefore braid groups are precisely the fundamental group of such spaces. Consequently, quantisation of a system of n indistinguishable, impenetrable particles must start from such highly complicated, multiply-connected space [LM77, RBH11, BR12, RH08, HBC20, RT20].

Wave functions for bosons and fermions are considered without resource to meaningless summations of wave functions of distinguishable particles, usually found in textbooks. Braid groups are not mentioned by name, but the role of the fundamental group is clearly stated. Since quantisation employs just the fundamental group of the configuration space [Sch68, LM71, Mor72, MMN79], then it must be involved with braid groups and, consequently, statistical mechanics follows suitably. Notice that impenetrability is an essential concept here. If strands could traverse each other, the fundamental group would be simply S_n. The origin of such hardcore feature is, for what concerns statistics, secondary. Particles governed by intermediate statistics, as given by the braid group, have been called *anyons*. We are avoiding such terminology only because people working on the subject have been mainly worried about the internal properties of such objects, for instance proposing models for their structure. Although this structure is supposed to be at the origin of the required impenetrability, we are here only interested in the consequences of this property, whatever its origin may be. Let us only retain that quantum mechanics of a system of impenetrable particles is regulated by B_n and not by S_n. Since braids can be unwoven in spaces of three or more dimensions (property formalised in the trivializing theorem below), then practical interest resides in two-dimensional systems. Of course, the interest in these statistics betides mainly on the possibility of their relationship with high-temperature superconductivity.

We might look at all this from still another point of view. The group of looping motions of n distinct interior points of a manifold M is just the pure braid group of M. There is much more, as braid groups in a sense classify motions of spaces in general, but we will not deal on such themes here forthwith.

1.8 More on the group structure

It was previously mentioned that just as S_n, the braid group B_n has $(n-1)$ generators, σ_j, satisfying relations (1.22, 1.23), similarly to relations (1.7, 1.8) for S_n. The absence of the condition (1.9), however, makes B_n a quite different group of infinite order, whereas S_n has finite order. In what follows a series of results are presented for a better understanding of relations between the braid and the symmetric group.

1.8.1 Braid group presentation

The symmetric group can be written as the quotient of the full braid group by the pure braid group. Besides, the braid group B_n is a fibered space. There exists a canonical isomorphism

$$B_n/P_n \simeq S_n, \tag{1.33}$$

where, recall, $P_n := \pi_1(F_nM)$. The inclusion mapping $\iota : F_nM^n \hookrightarrow M$ induces a homomorphism $\iota_* : \pi_k(F_nM) \to \pi_k(M)$, where π_k denotes the kth homotopy group.

Of great physical importance is the *trivializing theorem*:

☞ *If M is a closed smooth manifold, the mapping $\iota_* : \pi_k (F_n M) \to \pi_k (M)$ is surjective if $\dim M > k$, and also injective (and so an isomorphism) if $\dim M > k + 1$.*

For $M = S^3$, $\pi_1(M) = 1$ and $B_n = S_n$. The classes of loops, highly non-trivial on a punctured $\mathbb{E}^2$, become trivial in $\mathbb{E}^3$. In fact, each loop can be continuously deformed into a point in a punctured $\mathbb{E}^3$. This corresponds to the fact, mentioned in the preface, that braids can be unwoven in $\mathbb{E}^3$. Consequently, for $\dim M \geq 3$, statistical mechanics is just, as usual, regulated by the symmetric group, with bosons and fermions. But not so, for $\dim M = 2$. A gas in two dimensions has a different statistics, a *braid statistics*. This might be the case, for example, for the electron gas in superconductors, in which the electrons (by the Meissner effect) are confined to the surface.

We previously observed that $P_n = \ker h$, given by Eq. (1.21). To dig a little more about the pure braid group, let us introduce a presentation of P_n, by the generators

$$A_{ij} = \sigma_{j-1}\sigma_{j-2}\ldots\sigma_{i+1}\sigma_i^2\sigma_{i+1}^{-1}\ldots\sigma_{j-2}^{-1}\sigma_{j-1}^{-1}, \qquad (1.34)$$

and the awkward defining relations for the operation $A_{rs}A_{ij}A_{rs}^{-1}$:

$$
\begin{cases}
A_{ij}, & \text{if } \ r < s < i < j \quad \text{or} \quad i < r < s < j, \\
A_{rj}A_{ij}A_{rj}^{-1}, & \text{if } \ s = i, \\
A_{ij}A_{sj}A_{ij}A_{ij}A_{rs}^{-1}, & \text{if } \ r = i < j < s, \\
A_{rj}A_{sj}A_{rj}^{-1}A_{sj}^{-1}A_{ij}A_{rj}A_{sj}A_{rj}^{-1}A_{sj}^{-1}, & \text{if } \ r < i < s < j.
\end{cases}
$$

From these expressions for the generators, the meaning of A_{ij} can be understood by drawing

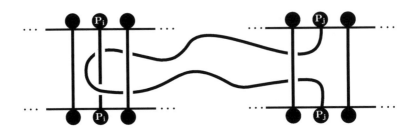

Figure 14: Representation of the A_{ij} in Eq. (1.34).

or, still better, by making experiments with real strings. For $n \geq$ 3, the center of B_n is infinite, cyclic, and generated by

$$(\sigma_1\sigma_2\sigma_3 \ldots \sigma_{n-1})^n = (A_{12})(A_{13}A_{23}) \ldots (A_{1n}A_{2n} \ldots A_{(n-1)n}). \quad (1.35)$$

1.9 The Burau representation

Things are sometimes easier to visualise in terms of matrices. Not only for this reason, but also for their importance for the knot invariants to be regarded later, let us examine a representation of B_n found by Burau in 1936 [Bur36]. One starts with a representation of the group S_n, precisely that given at the end of Sect. 1.2. Each basis matrix s_i' is diagonal, up to the ith row, whose $(i-1)^{\text{th}}$ and i^{th} entries are, respectively, 1 and -1. The Burau matrices are obtained by changing such entries, respectively, to t and $-t$. For $n = 4$, one gets the new matrices

$$\sigma_1 = \begin{pmatrix} 1 & 0 & 0 & 0 & 0 \\ t & -t & 1 & 0 & 0 \\ 0 & 0 & 1 & 0 & 0 \\ 0 & 0 & 0 & 1 & 0 \\ 0 & 0 & 0 & 0 & 1 \end{pmatrix}, \tag{1.36}$$

$$\sigma_2 = \begin{pmatrix} 1 & 0 & 0 & 0 & 0 \\ 0 & 1 & 0 & 0 & 0 \\ 0 & t & -t & 1 & 0 \\ 0 & 0 & 0 & 1 & 0 \\ 0 & 0 & 0 & 0 & 1 \end{pmatrix}, \tag{1.37}$$

$$\sigma_3 = \begin{pmatrix} 1 & 0 & 0 & 0 & 0 \\ 0 & 1 & 0 & 0 & 0 \\ 0 & 0 & 1 & 0 & 0 \\ 0 & 0 & t & -t & 1 \\ 0 & 0 & 0 & 0 & 1 \end{pmatrix}. \tag{1.38}$$

They satisfy the defining relations for B_4. The main novelty with respect to the case S_4 is that now $\sigma_k^2 \neq \mathbf{I}$, where $\mathbf{I}$ denotes the identity matrix. Indeed, their eigenvalues are $(-t)$ and 1, and it follows that $(\sigma_k - \mathbf{I})(\sigma_k + t\mathbf{I}) = 0$. This means that the squared generators are linear functions of the generators. Algebras with this property are called *Hecke algebras*, referring to an algebra which consists of the endomorphisms of a permutation representation of a topological group. The generators (1.38) satisfy the relations

$$\sigma_i \sigma_{i+1} \sigma_i = \sigma_{i+1} \sigma_i \sigma_{i+1}, \tag{1.39}$$

$$\sigma_i \sigma_j = \sigma_j \sigma_i, \qquad \text{for } |i - j| \geq 2, \tag{1.40}$$

$$\sigma_k^2 = (1 - t)\sigma_k + t\mathbf{I}. \tag{1.41}$$

Therefore, the Burau definition represents B_n on a Hecke algebra. The symmetric group is recovered when $t = 1$. S_n is a very particular case, where as a consequence of $s_k^2 = \mathbf{I}$, the group has finite order. A Hecke algebra is a kind of deformation of the S_n algebra, but its character is deeply different. A quick test with the matrices (1.38) convinces the reader that successive powers of each σ_k provide different matrices, so that the elements of B_n, which are the words formed by products of such powers, are infinite in number. The braid group B_n has, of course, infinite order, and its associated words are represented by the corresponding product of matrices.

Besides, the representation (1.38) is reducible. Still another representation comes out if one simply skips the first and the last rows and columns, getting so an $(n - 1) \times (n - 1)$ matrix representation. For $n = 4$,

$$\sigma_1 = \begin{pmatrix} -t & 1 & 0 \\ 0 & 1 & 0 \\ 0 & 0 & 1 \end{pmatrix}, \tag{1.42}$$

$$\sigma_2 = \begin{pmatrix} 1 & 0 & 0 \\ t & -t & 1 \\ 0 & 0 & 1 \end{pmatrix}, \tag{1.43}$$

$$\sigma_3 = \begin{pmatrix} 1 & 0 & 0 \\ 0 & 1 & 0 \\ 0 & t & -t \end{pmatrix} \tag{1.44}$$

is called the *reduced Burau representation*. It will be instrumental in getting Alexander's invariant polynomials for knots.

Recall that the braid group of n strands can be presented as

$$B_n = \left\langle \sigma_1, \ldots, \sigma_{n-1}, \quad \begin{matrix} \sigma_i \sigma_j = \sigma_j \sigma_i, & \text{for } |i - j| \geq 2 \\ \sigma_i \sigma_j \sigma_i = \sigma_j \sigma_i \sigma_j, & \text{for } |i - j| = 1 \end{matrix} \right\rangle, \tag{1.45}$$

where the braid σ_i performs a positive half-twist of the strands i and $i + 1$.

As pointed out by Majid, some representations of the braid group are not one-dimensional, necessarily [Maj93, Maj95]. Suppose that V is a vector space and take a representation of B_n on the tensor product space $V^{\otimes n} = V \otimes V \otimes V \otimes \cdots \otimes V$ involving n copies of V. Introduce a mapping from the ith elementary braid s_i to the linear mapping

$$v_1 \otimes v_2 \otimes \cdots \otimes v_n \mapsto v_1 \otimes \ldots \otimes R(v_i \otimes v_{i+1}) \otimes \cdots \otimes v_n, \tag{1.46}$$

where $v_k \in V$, $k \in \{1, \ldots, n\}$, and $R : V \otimes V \to V \otimes V$ is an invertible and linear operator representing a transposition between the ith and the $(i+1)^{\text{th}}$ entries in $V^{\otimes n}$, satisfying the Yang-Baxter equation

$$(R \otimes \text{id}) \circ (\text{id} \otimes R) \circ (R \otimes \text{id}) = (\text{id} \otimes R) \circ (R \otimes \text{id}) \circ (\text{id} \otimes R), \quad (1.47)$$

where id is the identity of V and "$\circ$" denotes the composition of operators in $V \otimes V$.

Solutions to the Yang-Baxter equations are closely related to quantum groups. Naturally, Hecke algebras can be introduced from the Yang-Baxter equation, if V is the linear span of two vectors $u, v \in V$, and R is defined by

$$
\begin{align}
R(u \otimes u) &= u \otimes u, & (1.48)\\
R(v \otimes v) &= v \otimes v, & (1.49)\\
R(u \otimes v) &= t(v \otimes u), & (1.50)\\
R(v \otimes u) &= t(u \otimes v) + (1 - t^2)(v \otimes u). & (1.51)
\end{align}
$$

Besides the operator R satisfies the Yang-Baxter equations, the relation

$$R^2 + (t^2 - 1)R - t^2 \, \text{id} = 0 \qquad (1.52)$$

also holds, and a Hecke relation is obtained.

1.10 Direct product representation

Given two vector spaces V and W, tensor products of linear mappings $A \in \text{End}(V)$[1] and $B \in \text{End}(W)$ can be introduced, by

---

[1]Hereon $\text{End}(V)$ denotes the space of endomorphisms of V.

defining a bilinear mapping $A \otimes B \in \mathrm{End}(V \otimes W)$. Given $v \in V$ and $w \in W$, it is defined as

$$(A \otimes B)(v \otimes w) = (A(v)) \otimes (B(w)) \tag{1.53}$$

The function $A \otimes B$ is named the tensor product of A and B.

⇨ **Comment 1.1.** Let $A = \begin{pmatrix} a & b \\ c & d \end{pmatrix}$ and $A' = \begin{pmatrix} a' & b' \\ c' & d' \end{pmatrix}$ matrices in $M(2, \mathbb{K})$, where $\mathbb{K}$ is a general field. The function $A \otimes A' \in \mathrm{Hom}(\mathbb{K}^2 \otimes \mathbb{K}^2)$ is bilinear. Now, denote by $\{e_1, e_2\}$ the canonical basis of $\mathbb{K}^2$. The matrix of $A \otimes A'$ in the basis $\{e_1 \otimes e_1, e_1 \otimes e_2, e_2 \otimes e_1, e_2 \otimes e_2\}$ can be immediately calculated. By definition,

$$
\begin{aligned}
(A \otimes A')(e_1 \otimes e_1) &= A e_1 \otimes A' e_1 \\
&= (a e_1 + c e_2) \otimes (a' e_1 + c' e_2) \\
&= a a' e_1 \otimes e_1 + a c' e_1 \otimes e_2 + c a' e_2 \otimes e_1 + c c' e_2 \otimes e_2, \\
(A \otimes A')(e_1 \otimes e_2) &= A e_1 \otimes A' e_2 \\
&= (a e_1 + c e_2) \otimes (b' e_1 + d' e_2) \\
&= a b' e_1 \otimes e_1 + a d' e_1 \otimes e_2 + c b' e_2 \otimes e_1 + c d' e_2 \otimes e_2, \\
(A \otimes A')(e_2 \otimes e_1) &= A e_2 \otimes A' e_1 \\
&= (b e_1 + d e_2) \otimes (a' e_1 + c' e_2) \\
&= b a' e_1 \otimes e_1 + b c' e_1 \otimes e_2 + d a' e_2 \otimes e_1 + d c' e_2 \otimes e_2, \\
(A \otimes A')(e_2 \otimes e_2) &= A e_2 \otimes A' e_2 \\
&= (b e_1 + d e_2) \otimes (b' e_1 + d' e_2) \\
&= b b' e_1 \otimes e_1 + b d' e_1 \otimes e_2 + d b' e_2 \otimes e_1 + d d' e_2 \otimes e_2.
\end{aligned}
$$

It means that the matrix representation associated with $A \otimes A'$ is given by

$$
\begin{pmatrix}
aa' & ab' & ba' & bb' \\
ac' & ad' & bc' & bd' \\
ca' & cb' & da' & db' \\
cc' & bd' & db' & dd'
\end{pmatrix}
= \begin{pmatrix} aA' & bA' \\ cA' & dA' \end{pmatrix}. \tag{1.54}
$$

In this example the trace of the tensor product representation is given by

$$
\begin{aligned}
\mathrm{Tr}\,(A \otimes A') &= a(a' + d') + d(a' + d') = (a + d)(a' + d') \\
&= (\mathrm{Tr}\,A)(\mathrm{Tr}\,A'). \tag{1.55}
\end{aligned}
$$

✓

⇨ **Comment 1.2.** Consider $[a_{ij}]$ the entries of a matrix A, with respect to the basis $\{e_1, \ldots, e_n\} \subset V$ and $[b_{kl}]$ the matrix of the operator B with respect to the basis $\{f_1, \ldots, f_m\} \subset W$. By the same procedure illustrated in the previous example, the matrix representation associated with $A \otimes B$, with respect to the basis $\{e_1 \otimes f_1, e_1 \otimes f_2, \ldots, e_1 \otimes f_m, e_1 \otimes f_1, e_2 \otimes f_2, \ldots, e_2 \otimes f_m, \ldots, e_n \otimes f_m\} \subset V \otimes W$, is provided by

$$A \otimes B = \begin{pmatrix} a_{11}B & a_{12}B & \cdots & a_{1n}B \\ a_{21}B & a_{22}B & \cdots & a_{2n}B \\ \vdots & \vdots & \ddots & \vdots \\ a_{n1}B & a_{n2}B & \cdots & a_{nn}B \end{pmatrix}. \tag{1.56}$$

This matrix is denominated the *Kronecker product* of A and B. ✓

Other representations of the braid groups are obtained with the use of direct products of matrix algebras. When $V = W$, given the direct product $A \otimes B$ of two square matrices A and B, the bracket notation yields

$$\langle ij|A \otimes B|mn \rangle = \langle i|A|m \rangle \langle j|B|n \rangle \tag{1.57}$$

The notation $|mn\rangle$ is equivalent to the tensor product $|m\rangle \otimes |n\rangle$ between vectors $|n\rangle, |m\rangle \in V$, whereas $\langle n|$ denotes vectors in the dual space V^*. Similarly, the direct product of three matrices is given by

$$\langle ijk|A \otimes B \otimes C|mnr \rangle = \langle i|A|m \rangle \langle j|B|n \rangle \langle k|C|r \rangle. \tag{1.58}$$

The direct product notation make expressions to be more compact, in the following way. Let $T = A \otimes B$, and I be the identity matrix. Hence one writes

$$T_{12} = A \otimes B \otimes I, \tag{1.59}$$
$$T_{13} = A \otimes I \otimes B, \tag{1.60}$$
$$T_{23} = I \otimes A \otimes B. \tag{1.61}$$

A useful property of direct products of matrices is given by

$$(A \otimes B \otimes C)(G \otimes H \otimes J) = (AG) \otimes (BH) \otimes (CJ), \quad (1.62)$$

and analogously for higher order products. One can also use the notation

$$T^{ij}{}_{mn} = \langle ij|T|mn \rangle. \quad (1.63)$$

Given a matrix R, an expression like

$$R^{kj}{}_{ab} R^{bi}{}_{cr} R^{ac}{}_{mn} = R^{ji}{}_{ca} R^{kc}{}_{mb} R^{ba}{}_{nr} \quad (1.64)$$

is equivalent to

$$R_{12} R_{23} R_{12} = R_{23} R_{12} R_{23}, \quad (1.65)$$

which is the braid equation. To evince it, look at s_1, s_2 as $s_1 = S_{12}$ and $s_2 = S_{23}$, where S denotes some direct product as above. Then find

$$\langle ijk|s_1 s_2 s_1|mnr \rangle = S^{ij}{}_{pq} S^{qk}{}_{vr} S^{pv}{}_{mn}, \quad (1.66)$$

$$\langle kji|s_2 s_1 s_2|mnr \rangle = S^{ji}{}_{qs} S^{kq}{}_{mv} S^{vs}{}_{nr}, \quad (1.67)$$

so that the braid equation reads

$$S^{kj}{}_{ab} S^{bi}{}_{cr} S^{ac}{}_{mn} = S^{ji}{}_{ca} S^{kc}{}_{mb} S^{ba}{}_{nr}. \quad (1.68)$$

Without going into the detailed matrix elements calculations, we can exhibit a strong relationship of braid relations with projectors. Let one defines

$$s_1 = A \otimes B \otimes I, \qquad s_2 = I \otimes A \otimes B. \quad (1.69)$$

The operators A and B can be seen as automorphisms of some vector space V ($A, B \in \text{Aut}(V)$ and $A \otimes B \in \text{Aut}(V \otimes V)$). Therefore

$$s_1 s_2 s_1 = A^2 \otimes BAB \otimes B, \quad (1.70)$$

$$s_2 s_1 s_2 = A \otimes ABA \otimes B^2, \quad (1.71)$$

so that the braid relations require that A and B be projectors $(A^2 = A, \ B^2 = B)$ satisfying the relation $ABA = BAB$. It follows that $s_1^2 = s_1, s_2^2 = s_2$. We will later find more involved sets of projectors.

So far we have found conditions for representations of the braid group B_3. Higher order direct products of projectors will produce representations for higher order braid groups. In the general case, given a matrix $R \in \text{Aut}(V \otimes V)$ (the automorphism space of $V \otimes V$) satisfying relations as above and the identity $I \in \text{Aut}(V)$, a representation of B_n on $V^{\otimes n}$ is obtained with generators

$$\sigma_i = I \otimes I \otimes \cdots \otimes R_{i,i+1} \otimes \cdots \otimes I \otimes I = I^{\otimes(i-1)} R_{i,i+1} I^{\otimes(n-i)}. \quad (1.72)$$

1.11 Relation to Yang-Baxter equation

With the notation previously fixed, one can easily establish a direct connection relating the braid relations to the Yang-Baxter equation, usually written as

$$R_{12} R_{13} R_{23} = R_{23} R_{13} R_{12}, \quad (1.73)$$

which is the same as

$$R^{jk}{}_{ab} R^{ib}{}_{cr} R^{ca}{}_{mn} = R^{ij}{}_{ca} R^{ck}{}_{mb} R^{ab}{}_{nr}. \quad (1.74)$$

Define now another product matrix by the permutation $\widehat{R} = PR$, in such a way that

$$\widehat{R}^{ij}{}_{mn} = R^{ji}{}_{mn}. \quad (1.75)$$

Therefore Eq. (1.73) reads

$$\widehat{R}_{12} \widehat{R}_{23} \widehat{R}_{12} = \widehat{R}_{23} \widehat{R}_{12} \widehat{R}_{23}, \quad (1.76)$$

which is precisely the braid equation. The permutation relation is thus a very interesting tool to obtain braid group representations from Yang-Baxter solutions and vice-versa. Notice that, due to this equivalence, many people give the name Yang-Baxter equation to the braid equation itself. An important point is that Yang-Baxter equations come out naturally from the representations of the Lie algebra of any Lie group. Thus, each Lie algebra representation provides a solution of the braid relations [YG89].

The relation between this matrix formulation and our first informal representation of braids by their plane drawings leads to an instructive matrix-diagrammatic formulation. It is enough to notice the relationship

$$
\begin{array}{ccc}
a \qquad b & & \\
\diagup\!\!\!\!\!\diagdown & \leftrightarrow & \hat{R}^{ab}{}_{cd} \\
c \qquad d & &
\end{array}
\tag{1.77}
$$

and proceed to algebrise diagrams by replacing concatenation by matrix multiplication, paying due attention to the contracted indexes. Looking at Fig. 15, one sees that the braid equation becomes exactly the Yang-Baxter equation in its form (1.64).

Kerner [Ker00] described a ternary aspect of these R-operators discovered by Okubo in the search for new solutions of Yang-Baxter equations [Oku931, Oku932], introducing a parameter $\theta \in \mathbb{R}$, in such a way that

$$
R^{ab}{}_{cd}(\theta)R^{ef}{}_{bg}(\theta')R^{hi}{}_{ae}(\theta'') = R^{ea}{}_{cg}(\theta'')R^{hb}{}_{ce}(\theta')R^{if}{}_{ba}(\theta) \tag{1.78}
$$

holds, with $\theta' = \theta + \theta''$. Okubo then considered the following symplectic and orthogonal vector spaces endowed simultaneously with a non-degenerate bilinear form

$$
\langle u, v \rangle : V \otimes V \to \mathbb{C}, \qquad u, v \in V, \tag{1.79}
$$

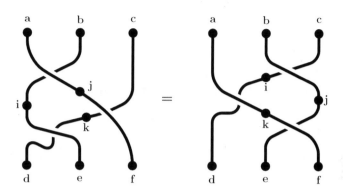

Figure 15: The relation $R^{ab}{}_{ij} R^{jc}{}_{kf} R^{ik}{}_{de} = R^{bc}{}_{ij} R^{ai}{}_{dk} R^{kj}{}_{ef}$.

and a triple product

$$\{u, v, w\} : V \otimes V \otimes V \to V, \ u, v, w \in V. \qquad (1.80)$$

These products are defined by the following relations:

a) $\langle v, u \rangle = \varepsilon \langle u, v \rangle$,

b) $\{v, u, w\} = -\varepsilon \{u, v, w\}$.

If $\varepsilon = -1$, the system is called symplectic and if $\varepsilon = 1$, it is called orthogonal. Besides, the following relations can be verified:

c) $\langle \{u, v, w\}, z \rangle = -\langle w, \{u, v, z\} \rangle$,

d) $\{u, v\{x, y, z\}\} = \{\{u, v, x\}, y, z\} + \{x, \{u, v, y\}, z\} + \{x, y, \{u, v, z\}\}$,

e) $\{u, v, w\} + \varepsilon \{u, w, v\} = 2\alpha \langle v, w \rangle u - \alpha \langle u, v \rangle w - \alpha \langle w, u \rangle v$, for some $\alpha \in \mathbb{R}$.

Using a basis $\{e_a\}$ of V, the bilinear form presented in Eq. (1.79) can be written as $\langle e_i, e_k \rangle = g_{ik} = \varepsilon\, g_{ki}$, together with the triple product $\{e_i, e_k, e_m\} = C^j{}_{ikm} e_j$, where the coefficients $C^j{}_{ikm}$ are ternary structure constants.

If a one-parameter family of triple products is defined on a basis $\{e_i, e_k, e_m\}_\theta$ of V, then a R-operator also depends on the same parameter θ, as

$$R^{ij}{}_{km} = \langle e^i \{e^j, e_k, e_m\}_\theta, \rangle \tag{1.81}$$

or equivalently,

$$\{e^b, e_c, e_d\}_\theta = R^{ab}{}_{cd} e_a. \tag{1.82}$$

where $e^a = g^{ab} e_b$ [VR16]. The symmetry condition $R^{ba}{}_{dc}(\theta) = R^{ab}{}_{cd}(\theta)$ can be now written as

$$\langle u, \{v, w, z\}_\theta \rangle = \langle z, \{w, v, u\}_\theta \rangle \tag{1.83}$$

and the Yang-Baxter equation becomes equivalent, with an extra condition imposed on the ternary product:

$$\{v, \{u, e_a, z\}_{\theta'}, \{e^a, x, y\}_\theta\}_{\theta''} = \{u, \{v, e_a, x\}_{\theta'}, \{e^a, z, y\}_{\theta''}\}_\theta \tag{1.84}$$

Okubo found new solutions of Yang-Baxter equations in terms of one-parameter families of ternary products satisfying the constraints (1.84) [Ker00, Oku931, Oku932]. We do not want to go beyond this formalism since it is not the central idea of this book. We will address more details in the next chapters.

Exercises

(1) Let U, V and W be vector spaces over the same field $\mathbb{K}$. Consider a basis $\{e_i\} \subset V$ and a basis $\{f_j\} \subset W$. Hence, the set $\{e_i \otimes f_j\}$ is a basis of $V \otimes W$. Although the tensor product between two vector spaces is not commutative, it is possible to establish an isomorphism

$$
\begin{aligned}
c_{V,W} : V \otimes W &\to W \otimes V \\
v \otimes w &\mapsto c_{V,W}(v \otimes w) := w \otimes v.
\end{aligned} \qquad (1.85)
$$

One can also require that a basis $\{e_i \otimes f_j\} \subset V \otimes W$ is mapped into the basis $\{f_j \otimes e_i\} \subset W \otimes V$, by the isomorphism (1.85).

a) Show that

$$
c_{V,W} \circ c_{W,V} = \mathrm{id}_{W \otimes V}, \qquad (1.86)
$$

and

$$
c_{W,V} \circ c_{V,W} = \mathrm{id}_{V \otimes W}. \qquad (1.87)
$$

b) Given the vector space $U \otimes V \otimes W$, show that

$$
\begin{aligned}
(c_{V,W} \otimes \mathrm{id}_U) &\circ (\mathrm{id}_V \otimes c_{U,W}) \circ (c_{U,V} \otimes \mathrm{id}_W) \\
&= (\mathrm{id}_W \otimes c_{U,V}) \circ (c_{U,W} \otimes \mathrm{id}_V) \circ (\mathrm{id}_U \otimes c_{V,W}). \quad (1.88)
\end{aligned}
$$

This equation is the Yang-Baxter equation.

(2) Discuss that any topological equivalence move possible on a knot can be reduced to Reidemeister moves.

(3) Eq. (1.25) displays the Artin presentation of the braid group. Show that denoting $\zeta = \sigma_1 \sigma_2 \ldots \sigma_{n-1}$ and $a = \sigma_1$, one can define the Coxeter presentation of the braid group:

$$
B_n = \left\langle \zeta, a \mid \zeta^n = (\zeta a)^{n-1}, \ a \zeta^{-i} a \zeta^i = \zeta^{-i} a \zeta^i a \right\rangle, \qquad (1.89)
$$

for $2 \leq i \leq \frac{n}{2}$. (As asserted in Ref. [Ban09], one of the advantages of the Coxeter presentation is the two-generator definition of braid groups, although some of the topological correspondence lacks when compared to the Artin presentation).

(4) Birman, Ko and Lee defined generators [BKL98]

$$b_{k\ell} = \left(\sigma_{k-1}\sigma_{k-2}\ldots\sigma_{\ell+1}\right)\sigma_\ell\left(\sigma_{\ell+1}^{-1}\sigma_{\ell+2}^{-1}\ldots\sigma_{k-1}^{-1}\right), \qquad (1.90)$$

that represent a crossing between two arbitrary braid strands k and ℓ. Regarding $b_{k\ell}$, the strand k overcrosses the strand ℓ and both strands overcross all other strands between them, to cross and therefore overcross the in-between strands again to return to their original positions [Ban09]. Show that using the generators (1.90), the so-called band-generator presentation of the braid group,

$$B_n = \langle\, a_{ts}, \quad 1 \le s < t \le n \rangle, \qquad (1.91)$$

with

$$a_{ts}a_{sr} = a_{tr}a_{ts} = a_{sr}a_{tr}, \qquad (1.92)$$
$$a_{ts}a_{rq} = a_{rq}a_{ts}, \quad \text{for} \quad (t-r)(t-q)(s-r)(s-q) > 0, \quad (1.93)$$

is obtained.

(5) Assume that at the bottom of the braid, one numbers the strands with labels $\{1,\ldots,n\}$, from left to right. Each crossing in which the strand i overcrosses the strand j is labelled by the generator g_{ij}. This defines the colored braid presentation [Ban09]. This presentation does not make explicit whether a crossing is positive or negative, in the Artin sense, also yielding $g_{ij}g_{ij} = \text{id}$. Show that the braid group relations in the colored braid presentation read

$$B_n = \langle\, g_{ij}, \quad \text{for} \quad i \ne j, \quad 1 \le i,j \le n \rangle, \qquad (1.94)$$

such that

$$g_{ij}g_{ik}g_{jk} = g_{jk}g_{ik}g_{ij},$$
$$g_{ij}g_{k\ell} = g_{k\ell}g_{ij}, \quad \text{for} \quad (i-k)(i-\ell)(j-k)(j-\ell) > 0. \qquad (1.95)$$

(6) One defines the fundamental braid word $\Delta_n \in B_n$ by

$$\Delta_n = \sigma_1\sigma_2 \ldots \sigma_{n-1}\sigma_1\sigma_2 \ldots \sigma_{n-2} \ldots \sigma_1\sigma_2\sigma_1. \qquad (1.96)$$

Show that taking into account the braid group relations yields

$$\Delta_n^2 = (\sigma_1\sigma_2 \ldots \sigma_{n-1})^n. \qquad (1.97)$$

(7) With the notation in exercises (5) and (6), show that in the colored braid presentation, the fundamental braid Δ_n reads [Ban09]

$$\begin{aligned} \Delta_n &= g_{12}g_{13} \cdots g_{1n}g_{23}g_{24} \cdots g_{n-1,n} \\ &= \prod_{i=1}^{n-1} \prod_{j=i+1}^{n} g_{ij}. \end{aligned} \qquad (1.98)$$

Chapter 2

Knots and links

Besides sailors and surgeons, perhaps one of the greatest specialists in knots has been the fisherman. In the past, the hangmen also occupied a side role among these professionals. To stimulate the reader, Fig. 16 shows the simplest example of their art. One may wonder about the falsity of appearances, in the case of the hangmen.

As anyone can find by experience with a string, in general two knots are non-equivalent. This means that one knot cannot be obtained from the other one without somehow tying or untying, namely, passing one of the ends through some loop. The mathematical formalisation conflating this intuitive notion of different knots is the so-called knot problem, and leads to an involved

Figure 16: A Solomon's seal knot, with braid word σ_1^5, also known as the cinquefoil knot.

theory. The characterisation is completely provided by the notion of knot- (or link-) type, which is as sound as unpractical. There is no practical way to establish the distinction between two given knots. However, two methods are allowing an imperfect solution for the problem. One of them attributes to every given knot (or link) a certain group, whereas the other one attributes a polynomial, instead. They are incomplete in the sense that two different knots can have the same polynomial or group. The characterisation by the knot groups is stronger and emulates the former one. Two knots with the same group have the same polynomials, but not vice-versa. On the other hand, polynomials are easier to be found out.

Links are intertwined knots, defined as smooth embeddings of spaces that are homeomorphic to a disjoint union of circles. A link is a collection of disjoint knots, possibly linked each other. Fig. 17 shows well-known examples: the simple and twofold links, the trefoil and the Borromean rings.

Figure 17: The simple and twofold links, the trefoil and the Borromean rings.

2.1 Essential fundamentals on topology

In this section, some basic topological concepts are going to be introduced, mainly focusing on the tools to be used in the subsequent chapters. More details can be checked, for instance, in Refs. [Ald01, AP95, Nak90, RV07].

2.1.1 Generalities

An equivalence relation R in a set X is a relation (i) *reflexive*, (ii) *symmetric*, and (iii) *transitive*, respectively meaning that (i) xRx, (ii) if xRy then yRx, and (iii) if xRy and yRz then xRz, $\forall x, y, z \in X$. The set of all elements that are equivalent to x constitutes an equivalence class of x, denoted by

$$[x] = \{y \in X \mid yRx\}. \tag{2.1}$$

The set of such (disjoint) equivalence classes is denoted by

$$\mathcal{X} = X/R = \{[x] \mid x \in X\} \tag{2.2}$$

and denominated the *quotient space*. A notation that is frequently used for xRy is $x \sim y$. An element $x \in [x]$ (or any another element in $[x]$) is a *representative* of $[x]$ [VR16].

⇨ **Comment 2.1.** Equivalence classes are disjoint sets: $[x] \cap [y] = \varnothing$. One can show that if $[x] \cap [y] \neq \varnothing$, then $[x] = [y]$. In fact, by supposing that $[x] \cap [y] \neq \varnothing$, there exists $c \in [x] \cap [y]$ such that $c \sim x$ and $c \sim y$. By transitivity, $x \sim y$. Now let us show that $[x] \subset [y]$: by taking an arbitrary element $x_1 \in [x]$ it follows that $x_1 \sim x$, and as $x \sim y$, thus $y \sim x_1$, $x_1 \in [y]$ therefore $[x] \subset [y]$. Analogously one can show that if $[y] \subset [x]$, then $[x] = [y]$ [VR16]. ✓

⇨ **Comment 2.2.** In the set of integers $\mathbb{Z}$ one can define the following equivalence relation:

$$m \sim n \Leftrightarrow m - n = kN, \quad k = 0, \pm 1, \pm 2, \ldots$$

namely, the integers m and n are considered equivalent if they differ by a multiple of N. The equivalence class of m consists therefore of the set

$$[m] = \{n \in \mathbb{Z} | n = m + kN\} = \{m, m \pm N, m \pm 2N, m \pm 3N, \ldots\}.$$

The set of all equivalence classes is the set $\mathbb{Z}_N = \mathbb{Z}/\sim$, the set of integers modulo N. Recall that when $a, b \in \mathbb{Z}$, the element a is said to be congruent to b mod n, denoted by $a \equiv b \,(\mathrm{mod}\ n)$, if $n|\,(a - b)$. One can show that in this case $a \equiv b \,(\mathrm{mod}\ n)$ if and only if there exists an integer k such that $a = b + kn$. One therefore defines the set of equivalence classes as $a \sim b \Leftrightarrow a = b \,(\mathrm{mod}\ n)$. The set denoted by

$$\mathbb{Z}_n = \{[0], [1], \ldots, [n - 1]\}, \qquad n \in \mathbb{N}, \quad n \geq 2, \tag{2.3}$$

can be endowed with a operation of sum, defined by

$$[a] \dotplus [b] \equiv (a + b) \,(\mathrm{mod}\ n) = [a + b], \qquad [a], [b] \in \mathbb{Z}_n. \tag{2.4}$$

A product operation $\bullet : \mathbb{Z}_n \times \mathbb{Z}_n \to \mathbb{Z}_n$ can be also defined by

$$[a] \bullet [b] \equiv ab \,(\mathrm{mod}\ n) = [ab], \tag{2.5}$$

where juxtaposition denotes the product in $\mathbb{Z}$. The elements of this group can be thought of as the congruence classes, also known as residues modulo n, that are coprime to n. Hence another name is the group of primitive residue classes modulo n. If the set $\mathbb{Z}_n$ is a field with the operations defined in Eqs. (2.4, 2.5), then n is a prime number. ✓

Going back to the topology, let A be a subset of a metric space. A point $p \in A$ is said to be an *interior point* of A when it is the center of an open ball. An open ball in A, of center a and radius r, is the set $B(a, r)$ of points in A whose distance to the point a is less than r. The *interior* of A is the set of interior points of A. When the open ball contains any point $q \in A$ that is not in A, the point q is said to be in the boundary of A. More precisely the boundary of A, denoted by ∂A, is the set of points which are the center of open balls that contain at least one point of A and one point of the complementary set of A. The subset A is said to be *closed* when $\partial A = \varnothing$.

A *topological space* consists of a set A and a collection P of subsets of A, denominated *open sets* of A, that satisfy: (i) $\varnothing$ and A are in P; (ii) the intersection of any subcollection of sets in P is in P; (iii) the union of any finite subcollection of sets in P is in P. P is called a topology, and the 2-tuple (A, P) is denominated a *topological space*. It is sometimes denoted simply by A. If A is a metric space, therefore (A, P) is called a Hausdorff space. If the complement C of a part of A is an open set, then C is a closed set in this topology. The closure of a part C in A is the least closed set of A containing C.

Compositions can be defined in topological spaces. For instance, the sum of topological spaces (A_1, P_1) and (A_2, P_2) consists of a collection of open sets $\{V_1 \cup V_2 \,|\, V_1 \subset P_1, V_2 \subset P_2\}$ related to the union $A_1 \cup A_2$. Moreover, the product of topological spaces (A_1, P_1) and (A_2, P_2) is defined as being the collection of open sets $\{V_1 \times V_2 \,|\, V_1 \subset P_1, V_2 \subset P_2\}$ related to the Cartesian product $A_1 \times A_2$.

The mapping $f : X \to Y$ between two topological spaces X and Y is said to be continuous if, for each open set $S' \subset Y$, its inverse image $f^{-1}(S')$ is an open set X. A homeomorphism is a continuous bijection $g : X \to Y$ whose inverse is also continuous. Two topological spaces are said to be topologically equivalent if there exists a homeomorphism between them. Properties that are preserved by homeomorphism are denominated topological properties. If the topological space X is a subset of $\mathbb{R}^n$, this last is called a *host space*.

$\Rightarrow$ **Comment 2.3.** The interval $X = (-\pi/2, \pi/2)$ is homeomorphic to the line $Y = \mathbb{R}$. Indeed, let us define $f : X \to Y$ given by $f(a) = \tan a$. The function $\tan a$ is one-to-one as well as its inverse $\arctan b$, for $a, b \in \mathbb{R}$. ✓

A topological space X is connected if it cannot be written as the union $X = X_1 \cup X_2$, where X_1 and X_2 are both open sets and $X_1 \cap X_2 = \varnothing$. Otherwise, X is called disconnected. For instance, the real line $\mathbb{R}$ is a connected metric space, and the metric space $\mathbb{R}\backslash\{0\}$ is not connected, since it can be expressed as the union of two disjoint non-empty open subsets. In fact, $\mathbb{R}\backslash\{0\} = (-\infty, 0) \cup (0, \infty)$. The connectedness of a topological space is independent of the space wherein it is embedded, being an intrinsic property. Some straightforward properties of topological spaces, whose demonstrations can be found at basic books of topology [Ald01, AP95, Nak90, RV07], are (i) the image of a connected set by a continuous function is a connected set, meaning that connectedness is a topological invariant; (ii) the closure of a connected set is connected; (iii) the Cartesian product of metric spaces is connected if and only if each metric space is connected.

A metric space X is named a *topological manifold* when, for all $p \in X$, there exists a neighborhood $D \subset X$ homeomorphic to an open subset V, namely, when there exists a homeomorphism $f : D \to V$.

Given two points a and b in a metric space $X \subset F$, a *path* between two points is a continuous mapping $f : [0, 1] \to X$, with $f(0) = a$ and $f(1) = b$. The set X is said to be *path connected* whenever two points in X can be linked by a path in X. A metric space is said to be locally path-connected when for any point $p \in X$ and every neighborhood D containing the point P, there exists a path-connected neighborhood V such that $p \in V \subset D$.

Denoting by L a set of indexes, a *covering* of a topological manifold X is a collection $X = \{C_\lambda\}_{\lambda \in L}$ of subsets F, such that $X \subset \cup_{\lambda \in L} C_\lambda$. If all subsets $\{C_\lambda\}_{\lambda \in L}$ are open, X is said to be an open covering. If there exists a subset $L_0 \subset L$, such that $X \subset \cup_{\lambda \in L_0} C_\lambda$, then a collection $\{C_\lambda\}_{\lambda \in L_0}$ is said to be a subcovering C. A topological space $X \subset F$ is *compact* if each open covering has a finite subcovering.

2.1.2 Homotopy

Given $X, Y \subset W$ topological spaces, two continuous mappings $f, g : X \to Y$ are said to be homotopic if, for all $x \in X$, there exists a continuous function $h : X \times [0, 1] \to W$, with $h(x, 0) = f(x)$ and $h(x, 1) = g(x)$. Such function is said to be an *homotopy*.

Homotopy is a fundamental topological concept which describes the equivalence between curves, surfaces, or, more generally, topological subspaces of a determined topological space, up to continuous deformations. It is straightforward to notice that homotopy is an equivalence relation, since a homotopy can be invertible, and two homotopies can be composed. When f and g are two curves (or paths) in $Y = \mathbb{R}^n$ defined on the same interval $X = [a, b]$, the homotopy h defines, for each parameter $t \in [0, 1]$, the curve $h(\cdot, t) : [a, b] \to \mathbb{R}^n$ that smoothly interpolates $h(\cdot, 0) = f$ and $h(\cdot, 1) = g$.

More generally, all closed curves in the plane are homotopic. Indeed, each curve can be contracted to a point, which is a very special case of a closed curve. A connected topological space which satisfies this property is called *simply connected topological space*. When the function $h(\cdot, t)$ is demanded to be a homeomorphism, for all values of $t \in [0, 1]$ under a deformation, there is a stronger concept: the *isotopy*. In fact, an isotopy of W is a continuous family of diffeomorphisms[1] ϕ_t of W, for $t \in [0, 1]$. For instance, smooth closed curves without self-interactions are in two isotopy classes, according to their orientation: clockwise or counterclockwise.

One can define a *loop* with base point x_0 as being a path in W with $f(0) = f(1)$ or, equivalently, a mapping from the 1-sphere S^1 into W. The notion of homotopy allows us to analyze a topological space X in terms of the juxtaposition of the representatives called *homotopic loops*, with a fixed point. The collection of these representatives, endowed with the juxtaposition operation, constitutes the so-called *fundamental group* of X. An important result for the classification of n-dimensional closed manifolds is that they are homeomorphic if and only if their fundamental groups are isomorphic. It means that the homeomorphism class of a closed manifold is completely determined by the class of isomorphism of the fundamental group. Details can be found in [Nak90].

Therefore one can show, with respect to the torus knot $T_{p,q}$, with particular cases illustrated in Fig. 4, that the fundamental group of their complements is given by

$$\pi(T_{p,q}) = \left\langle \omega, \tau \mid \omega^p = \tau^q \right\rangle. \tag{2.6}$$

Homotopy is an equivalence relation between continuous functions. Such equivalence relation splits the set of functions from

---

[1]Given two manifolds M and N, a differentiable mapping $f : M \to N$ is called a diffeomorphism if it is a bijection and its inverse $f^{-1} : N \to M$ is differentiable as well. If these functions are r times continuously differentiable, f is called a C^r-diffeomorphism.

X to Y into disjoint disconnected sets, consisting of equivalence classes. These classes are called homotopy classes, and $[f]$ denotes the class of the path f. The composition $f \circ g$ of functions (paths) preserves homotopy, which is also an operation between classes: $[f \circ g] = [f] \circ [g]$.

A group structure can be now introduced in the set of paths. Consider a topological space, X, and two paths f and g, respectively between x_0 and x_1; and x_1 and x_2, in X. The composition (or product) between f and g is the path

$$h(t) = (f \star g)(t) = \begin{cases} f(2t), & \text{if } t \in [0, 1/2), \\ g(2t - 1), & \text{if } t \in [1/2, 1], \end{cases} \qquad (2.7)$$

which defines a curve from x_0 to x_2, whose first half is given by f and the second half is given by g. It can be shown that the product $\star$ of paths is associative [AP95].

Two paths f and g having the same initial and final points are said to be homotopic paths when there exists a continuous function $F : [0, 1] \times [0, 1] \to X$ such that for all $s, t \in [0, 1]$,

$$F(s, 0) = f(s), \qquad (2.8)$$
$$F(s, 1) = g(s), \qquad (2.9)$$
$$F(0, t) = f(0) = x_0, \qquad (2.10)$$
$$F(1, t) = f(1) = x_1. \qquad (2.11)$$

The function F is said to be a path homotopy between f and g, and represents a continuous deformation of the curve f to the curve g.

Denoting by X a topological space, let us define now:

(a) the elements $e_x : [0, 1] \to X$ such that $e_x([0, 1]) = x$; the path e_x is indeed a constant path in X, which plays that role of identity under the composition $\star$,

(b) the inverse elements $f^{-1}(t) = f(1 - t)$, in such a way that $[f] \star [f^{-1}] = [e_{x_0}]$ and $[f^{-1}] \star [f] = [e_{x_1}]$.

The space of paths endowed with the operation $\star$ restricted to the space of closed paths constitutes a group, denominated *fundamental group*.

The set formed by the loop homotopy classes, with base point $x_0 \in X$, is called the *fundamental group of X with respect to the base point x_0*, denoted by $\pi_1(X, x_0)$. It is also called the *first homotopy group* of X in x_0. If X is path connected, then $\pi_1(X, x_0)$ is independent of the point x_0: given another point $x_1 \in X$, the groups $\pi_1(X, x_0)$ and $\pi_1(X, x_1)$ are isomorphic. Indeed, let us consider a path γ in X with initial and final points respectively provided by $x_0 = \gamma(0)$ and $x_1 = \gamma(1)$, and define the function

$$
\begin{aligned}
\mathring{\gamma} : \pi_1(X, x_0) &\rightarrow \pi_1(X, x_1) \\
[f] &\mapsto \mathring{\gamma}([f]) = [\gamma^{-1}] \star [f] \star [\gamma]
\end{aligned}
\qquad (2.12)
$$

As $[f]$ is a class of loops with base point x_0, therefore $[\gamma^{-1}]\star[f]\star[\gamma]$ is a loop with base point x_1. Besides, $\mathring{\gamma}$ is a group isomorphism [AP95, Nak90]. If X is path connected, the fundamental group is denoted by $\pi_1(X)$.

2.1.3 Homology

To compute the homotopic loop classes is, in general, not trivial. Poincaré introduced the term *homology* to study the homeomorphism between manifolds. Instead of evaluating the number of homotopic loop classes, one splits the topological space X in more simple geometric figures and analyzes the relationship among them. The number of obstructions, that prevents any loop in a topological space to be continuously deformed into a point, corresponds to the number of cyclic sequences of submanifolds of dimension less than n, that are not boundaries of a submanifold of dimension $n + 1$. Those cyclic sequences are denominated *n-dimensional cycles*. A circle is, for instance, a closed one-dimensional manifold which is not a boundary of any two-dimensional manifold, whereas a disk is constituted by a two-dimensional manifold that has a closed one-dimensional manifold contouring it.

The above mentioned obstruction is called *genus* of the manifold. Informally, one can assert that the sets of n-dimensional cycles that do not constitute boundaries of $(n+1)$-dimensional manifolds form the so-called n-dimensional homology group. The rank of this group is known to be the n-dimensional *Betti number*, denoted by β^n. The Betti numbers are topological invariants, and Poincaré showed that they are related to the Euler character χ by the expression

$$\chi = \sum (-1)^n \beta^n. \tag{2.13}$$

This invariant was previously discussed in the 18$^\text{th}$ century by Euler, who demonstrated that the number of vertexes minus the number of edges plus the number of faces in convex polyhedra is an invariant — and it was the first topological character ever studied. For convex polyhedra, one has $\chi = 2$. Notwithstanding, the expression known as Euler formula does not hold for manifolds with holes, which have Euler character $\chi = 2 - 2a$, where a is the genus of the manifold.

Now our interest is to decide when there is a partition of a topological manifold into topological submanifolds. Poincaré introduced simplicial complexes, constituted by simplexes. A n-dimensional simplex is the convex closure of a set of $n+1$ linearly independent elements $\{v_0, \ldots, v_n\}$ in some vector space. It is worth to recall that the convex closure of a subset S in a vector space, given $v_i \in S$, is the set

$$\left\{ \sum_{i=1}^n a_i v_i \in V \,\middle|\, a_j \in \mathbb{R}^+ \text{ and } \sum_{i=1}^n a_i = 1 \right\}, \tag{2.14}$$

of $n+1$ vertexes $v_0, v_1, v_2, \ldots, v_n$, such that the set $\{v_1 - v_0, v_2 - v_0, \ldots, v_n - v_0\}$ is linearly independent.

Another technique to obtain the partition of a topological space is by using cell complexes, constituted by cells. A n-dimensional cell is a space homeomorphic to $\mathbb{R}^n$. A cell complex in a topological space X must satisfy the following conditions: (i) each n-dimensional cell is homeomorphic to $\mathbb{R}^n$ and its boundary is constituted by the union of cells having dimension up to $n-1$; (ii) the closure of each cell intersects solely a finite number of other cells; (iii) any space $A \subset X$ is closed if and only if the intersection of A with the closure of any another cell is also closed.

⇨ **Comment 2.4.** A cube can be represented as a complex, constituted by eight 0-dimensional cells (vertexes), twelve one-dimensional cells, (edges), and six two-dimensional cells (faces). Note that a p-dimensional cell can represent any set of points topologically equivalent to $\mathbb{R}^p$. It means that with merely four types of cells (those with dimensions 0, 1, 2, and 3) one can describe a great variety of geometric objects embeddable in $\mathbb{R}^3$. ✓

More precisely, let us define the underlying simplex structure. Consider a k-simplex, σ, which is the convex closure of a set A of $k+1$ independents elements $\{a_0, \ldots, a_k\} \subset \mathbb{R}^p$, where $p \geq k$. The set A is said to generate the simplex σ. A simplex generated by a subset $B \subset A$ is called *face* of σ. If τ is a face of σ, one denotes $\tau \leq \sigma$. A face $B \subset A$ is said to be a proper face if $\varnothing \neq B \neq A$. A 0-dimensional face is called vertex, and a one-dimensional face is denominated by edge. An orientation of a k-simplex σ is induced by an ordering of its vertexes, and denoted by $\langle a_0 \cdots a_k \rangle$. For any permutation π of $\{0, \ldots, k\}$, the orientation $\langle a_{\pi(0)} \cdots a_{\pi(k)} \rangle$ equals $(-1)^{\epsilon(\pi)} \langle a_0 \cdots a_k \rangle$, where the sign $\epsilon(\pi)$ corresponds to the number of transpositions of π, in such a way that to each simplex one can associate two distinct orientations. A simplex and a choice of its orientation are called an oriented simplex. When τ is a $(k-1)$-dimensional face of σ, obtained by omitting the vertex a_i, the induced orientation in τ is $(-1)^i \langle a_0 \cdots \hat{a}_i \cdots a_k \rangle$, where $\hat{a}_i$ denotes the omission of a_i.

A simplicial complex K is a finite set of simplexes in $\mathbb{R}^m$, such that: (i) if σ is a simplex of K and τ is a face of σ, therefore τ is a simplex of K; (ii) if σ and τ are simplexes of K, thus $\sigma \cap \tau = \varnothing$ or a common face between σ and τ. The dimension of K is the maximum among the dimension of its simplexes. The underlying space of K, denoted by $|K|$, is the union of all simplexes of K, endowed with the topology of $\mathbb{R}^m$. The i-skeleton of K, denoted by K^i, is the union of all simplexes of K which has a maximal dimension i. A subcomplex $J \subset K$ is a subset of K which is a simplicial complex. A triangularisation of a topological space X is a pair (K, h), such that K is a simplicial complex simplicial and h is a homeomorphism from the underlying space $|K|$ to X. The Euler character of a simplicial d-complex K, denoted by $\chi(K)$, is the number $\sum_{i=0}^{d}(-1)^i \alpha_i$, where α_i is the number of i-simplexes of K. An example of a simplicial simplex is a graph, which is a one-dimensional simplicial complex.

Intuitively, the sphere and the torus have different shapes, in the sense that these surfaces are not homeomorphic. The proof can be based on the Jordan curve theorem: consider a simple closed curve on the torus that cannot be disconnected on it. Indeed, let $T^2 \subset \mathbb{R}^3$ be the torus of radius R, obtained by the rotation of a circle of radius r around the z-axis, in the plane x, y, with center $(0, R, 0) \in \mathbb{R}^3$, where $R > r$. The torus T^2 is a differentiable manifold when one introduces a local coordinate system for all points of the torus. To accomplish it, consider the angles $\{(u, v) \,|\, 0 < u < 2\pi,\, 0 < v < 2\pi\}$ and the mapping

$$\phi : U \;\to\; \mathbb{R}^3$$

$$(u, v) \;\mapsto\; \phi(u, v) = \begin{pmatrix} (R - r\cos u)\cos v \\ (R - r\cos u)\sin v \\ r\sin u \end{pmatrix}. \qquad (2.15)$$

It is not difficult to verify that the mapping ϕ covers the torus, except for a meridian and a parallel circle. To assert that a "simple closed curve on the torus that cannot be disconnected on it" corresponds to consider, for instance, the circle $v = $ constant. Such curves, whose complement is connected, do exist. If there existed a homeomorphism from the torus in the sphere, the image of the curve should be a simple closed curve on the sphere. By the Jordan curve theorem, the complement of this curve is not connected. Since connectedness is preserved by homeomorphisms, the complements of the curves, respectively on the torus and the sphere, are not homeomorphic. This contradiction proves that the torus and the sphere are not homeomorphic. This demonstration merely asserts that the sphere is not homeomorphic to a closed surface with a genus different of zero. However, it cannot be used to show that a surface with genera greater than one is not homeomorphic to the torus. Homology provides a systematic technique for generalizing the previous argument to more general spaces. One presents below the simplicial homology related to simplicial complexes, adequate for computational aims.

Consider a finite simplicial complex K. A simplicial k-chain is a formal sum $\sum_j a_j\sigma_j$ over oriented k-simplexes σ_j in K, with coefficients a_j in the field of rational numbers $\mathbb{Q}$. In other words, it can be considered as a rational vector (a vector with rational coefficients) which has entries indexed by oriented k-simplexes of K. Besides, by definition, $-\sigma = (-1)\sigma$ is obtained from a simplex σ, inverting its orientation. With the obvious definition of the operations of sum and multiplication by scalars, the set of all simplicial k-chains constitutes a vector space, denoted by $C_k(K, \mathbb{Q})$, and denominated the vector space of simplicial k-chains of K. The dimension of this vector space equals the number of k-simplexes of K. Hence, the Euler character associated with a m-dimensional simplicial complex in K can be expressed as the alternating sum of the dimensions of the space of k-chains [RV07]:

$$\chi(K) = \sum_{i=0}^{m} (-1)^i \dim C_i(K, \mathbb{Q}). \qquad (2.16)$$

The boundary operator $\partial_i : C_i(K, \mathbb{Q}) \to C_{i-1}(K, \mathbb{Q})$ can be now defined. Taking a k-simplex $\sigma = \langle v_{j_0} \cdots v_{j_k} \rangle$, $k > 0$, one defines

$$\partial_i \sigma = \sum_{j=0}^{i} (-1)^j \langle v_{j_0} \cdots \hat{v}_{i_j} \cdots v_{i_k} \rangle, \qquad (2.17)$$

that can be extended by linearity. One also defines $C_{-1}(K, \mathbb{Q}) = 0$ and $\partial_0 \equiv 0$. The boundary operator is a linear function between vector spaces, and it is immediate to verify that $\partial_k \partial_{k+1} = 0$ [RV07].

⇨ **Comment 2.5.** With respect to Fig. 18, let us consider the 2-chain [RV07]

$$\alpha = \langle v_1 v_4 v_2 \rangle + \langle v_2 v_4 v_5 \rangle + \langle v_2 v_5 v_3 \rangle + \langle v_3 v_5 v_6 \rangle + \langle v_1 v_3 v_6 \rangle + \langle v_1 v_6 v_4 \rangle. (2.18)$$

Therefore $\partial_2 \alpha = \beta - \chi$, where

$$\beta = \langle v_4 v_5 \rangle + \langle v_5 v_6 \rangle - \langle v_4 v_6 \rangle \qquad (2.19)$$
$$\chi = \langle v_1 v_2 \rangle + \langle v_2 v_3 \rangle - \langle v_1 v_3 \rangle. \qquad (2.20)$$

Since $\partial_1 \beta = 0 = \partial_1 \chi$, then it follows that $\partial_1 \partial_2 \alpha = 0$ [RV07]. ✓

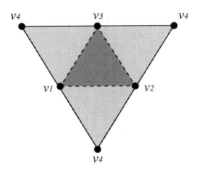

Figure 18: 1-chain and 2-chain in an *annulus*.

The vector space

$$Z_i(K, \mathbb{Q}) = \ker \partial_i \tag{2.21}$$

is called the space of simplicial k-cycles. The vector space

$$B_i(K, \mathbb{Q}) = \operatorname{Im} \partial_{i+1} \tag{2.22}$$

is denominated the vector space of simplicial k-boundaries. Since the boundary of a boundary is the empty set, $B_i(K, \mathbb{Q})$ is the subspace of $Z_i(K, \mathbb{Q})$. The quotient of vector spaces

$$H_i(K, \mathbb{Q}) = Z_i(K, \mathbb{Q})/B_i(K, \mathbb{Q}) \tag{2.23}$$

is the kth-homological vector space associated with K. In particular, two k-cycles α and β are k-homological if their difference is a k-boundary, namely, if there is a $(k + 1)$-chain ζ such that $\alpha - \beta = \partial_{k+1}\zeta$. The class of homology of $\alpha \in Z_i(K, \mathbb{Q})$ is indicated by $[\alpha]$. The ith Betti number associated with the simplicial complex K, denoted by $\beta^i(K, \mathbb{Q})$, is the dimension of $H_i(K, \mathbb{Q})$. In particular,

$$\beta^i(K) = \dim Z_i(K, \mathbb{Q}) - \dim B_i(K, \mathbb{Q}). \tag{2.24}$$

The coefficients of simplicial chains have been considered heretofore to be rationals, but more generally those coefficients can be elements of a ring, as the set of integers. In this case, one has a group structure, defining a group of homology. The Betti numbers $\beta^i(K)$ are the rank of such groups and indicate the number of i-dimensional genera in K [RV07].

The genus is one of the oldest topological invariants. Usually denoted by g, it is the maximum number of continuous closed curves that do not intersect, which can be represented on a surface without splitting it into two distinct domains. The genus associated with the sphere S^2 equals zero, whereas the torus has genus one. In general, the genus is related to the first Betti number by $2g = \beta_1$.

⇨ **Comment 2.6.** Consider a triangle of vertexes $\{v_1, v_2, v_3\}$. By the definition of boundary operators, $\partial_0(\langle v_i \rangle)$ (here $i = 1, 2, 3$), the kernel of ∂_0 is given by

$$\ker(\partial_0) = C_0 = \{c_1\langle v_1 \rangle + c_2\langle v_2 \rangle + c_3\langle v_3 \rangle \mid c_i \in \mathbb{Z}\} \simeq \mathbb{Z} \oplus \mathbb{Z} \oplus \mathbb{Z}.$$

Therefore all 0-chains are elements of $\ker(\partial_0)$ [RV07]. Now, for a 1-chain $\psi_1 = a_1\langle v_1 v_2 \rangle + a_2\langle v_2 v_3 \rangle + a_3\langle v_3 v_1 \rangle$, it follows that

$$\partial_1(\psi_1) = (a_3 - a_1)\langle v_1 \rangle + (a_1 - a_2)\langle v_2 \rangle + (a_2 - a_3)\langle v_3 \rangle \in \mathrm{Im}(\partial_1),$$

what means that the 0-chain $c_0 = c_1\langle v_1 \rangle + c_2\langle v_2 \rangle + c_3\langle v_3 \rangle$ is an element of the image of ∂_1 if and only if $c_1 = a_3 - a_1$, $c_2 = a_1 - a_2$, and $c_3 = a_2 - a_3$, having only two degrees of freedom for the choice of coefficients c_i. Hence

$$\mathrm{Im}(\partial_1) \simeq \mathbb{Z} \oplus \mathbb{Z}, \tag{2.25}$$

and therefore

$$H_0(K) = (\mathbb{Z} \oplus \mathbb{Z} \oplus \mathbb{Z})/(\mathbb{Z} \oplus \mathbb{Z}) \simeq \mathbb{Z}. \tag{2.26}$$

The computation of the other homology groups is more straightforward. As $\partial_1(\psi_1) = 0$ if and only if $a_1 = a_2 = a_3$, it follows that

$$\ker(\partial_1) = \{d\langle v_1 v_2 \rangle + d\langle v_2 v_3 \rangle + d\langle v_3 v_1 \rangle \mid d \in \mathbb{Z}\} \simeq \mathbb{Z}.$$

As there is no 2-chains, therefore $\ker(\partial_2) = \mathrm{Im}(\partial_2) = 0$, implying that

$$
\begin{aligned}
H_1(K) &= \ker(\partial_1)/\mathrm{Im}(\partial_2) = \ker(\partial_1) \simeq \mathbb{Z}, \\
H_2(K) &= \ker(\partial_2)/\mathrm{Im}(\partial_3) \simeq \{0\}.
\end{aligned}
$$

✓

Let us analyze some examples, illustrating the computation of the Betti numbers directly from the definition.

⇨ **Comment 2.7.** The Betti numbers associated with S^2 can be calculated as follows [RV07]. The simplicial complex K in Fig. 19 is the limit of

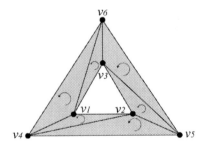

Figure 19: Calculations of the Betti numbers for S^2.

a 3-simplex, which consists of four 2-simplexes, six 1-simplexes and four 0-simplexes. For the sake of convenience, it is shown projected onto the plane, after extracting the boundaries incident to the 0-simplex v_4. The underlying space $|K|$ is homeomorphic to the two-dimensional sphere. Besides, vertexes with the same label must be identified, as the edges between the vertexes with the same label. The matrix of the boundary operator ∂_1 with respect to the canonical bases of $C_0(K,\mathbb{Q})$ and $C_1(K,\mathbb{Q})$ is provided by [RV07]

∂_1	$\langle v_1 v_2 \rangle$	$\langle v_1 v_3 \rangle$	$\langle v_1 v_4 \rangle$	$\langle v_2 v_3 \rangle$	$\langle v_2 v_4 \rangle$	$\langle v_3 v_4 \rangle$
$\langle v_1 \rangle$	-1	-1	-1	0	0	0
$\langle v_2 \rangle$	1	0	0	-1	-1	0
$\langle v_3 \rangle$	0	1	0	1	0	1
$\langle v_4 \rangle$	0	0	1	0	1	1

It follows that $\dim \mathrm{Im}(\partial_2) = 3$ and $\dim \ker(\partial_2) = 1$. Combining the previous results, one concludes that $\beta_0(K, \mathbb{Q}) = 1$, $\beta_1(K, \mathbb{Q}) = 0$ and $\beta_2(K, \mathbb{Q}) = 1$. ✓

Finally, the Euler character and the Betti numbers are intimately related. First, it is worth to emphasise that the Betti numbers are invariant under homotopy. In fact, if K_1 and K_2 are simplicial complexes with underlying spaces equivalent by homotopy, the ith homological vector spaces of K_1 and K_2 are isomorphic. In particular, $\beta_i(K_1, \mathbb{Q}) = \beta_i(K_2, \mathbb{Q})$, for all i.

Now, given a n-dimensional simplicial complex K, thus

$$\chi(K) = \sum_{i=0}^{n} (-1)^i \beta_i(K, \mathbb{Q}). \tag{2.27}$$

In fact, $\chi(K) = \sum_{i=0}^{d} (-1)^i \dim C_i(K, \mathbb{Q})$ and, as

$$H_i(K, \mathbb{Q}) = \ker(\partial_i)/\mathrm{Im}(\partial_{i+1}), \tag{2.28}$$

it follows that

$$\begin{aligned} \beta_i(K, \mathbb{Q}) &= \dim H_i(K, \mathbb{Q}) \\ &= \dim \ker(\partial_i) - \dim \mathrm{Im}(\partial_{i+1}) \\ &= \dim C_i(K, \mathbb{Q}) - \dim \mathrm{Im}(\partial_i) - \dim \mathrm{Im}(\partial_{i+1}), \end{aligned}$$

implying that $\chi(K) = \sum_{i=0}^{n} (-1)^i \beta_i(K, \mathbb{Q})$, since [RV07]

$$\sum_{i=0}^{n} (-1)^i \dim \mathrm{Im}(\partial_i) - \dim \mathrm{Im}(\partial_{i+1}) = 0. \tag{2.29}$$

The Betti numbers are going to be used further in Chapter 4.

2.2 Knot and link types, and knot group

Previously the continuous deformation of the host space, taking one knot into another was introduced. Now, this idea will be more precisely presented. False links, whose component knots are independent unknots, are similarly called unknots. Now, the formal definition:

> ☞ *A knot is a smooth embedding of S^1 into $\mathbb{R}^3$. Two knots K_1 and K_2 are said to be equivalent if there is an isotopy of $\mathbb{R}^3$ leading K_1 to K_2.*

Notice that, as spaces, all knots are topologically equivalent. Two knots are equivalent if one can be deformed into the other without ungluing the ends. Now comes the great question. How to characterise the difference between knots in general? How to put such a vague notion of difference between knots in a rigorous, formal, mathematically tractable form? The answer comes from noticing that tying and untying are performed in $\mathbb{E}^3$, and the equivalence or not is a consequence of how the circle is plunged in $\mathbb{E}^3$. The equivalence just introduced would not take into account different orientations of the knots. A more involved notion allows the knots K_1 and K_2 to be equal only if their orientations also coincide, once K_1 is deformed into K_2. Besides, arrows can be attached along the lines, to denote knots with an orientation.

We have already seen in the previous section that isotopies provide the finest notion of continuous orientation-preserving deformations of a space into itself. It is a special kind of homotopy, in which each member of the one-parameter family of deformations is invertible. This definition establishes an equivalence relation, whose classes are called *knot-types*. Knots that are equivalent to the circle itself are called *trivial* knots. The trefoil and the 4-knot have different and non-trivial types. In fact, it is equivalent (and in general more convenient) to use the sphere S^3 instead of the Euclidean space $\mathbb{E}^3$ as the host space. When a topological manifold M is the host space of links, this definition establishes an equivalence relation, whose classes are called *link-types*. Link-types provide the complete characterisation of links, but it has a serious drawback. Given a link, it is a very difficult task to perform isotopies to verify whether or not it can be taken into another given link. That is why sometimes the experts content themselves with incomplete characterisations, such as the link group and the polynomial invariants.

Going back to the braids discussed in Chapter 1, one can now formalise the concept of crossings, therein already discussed. A projection of a knot K is defined as the image of K under a projection $P : \mathbb{R}^3 \rightarrow \mathbb{R}^2$, such that the preimage of each point in $P(K)$ contains at most two points and there are (finite) many points in $P(K)$ with two preimages. Such points are called *crossings* and embody the concepts of *over* and *under* strands of general braids. In other words, a two-dimensional picture of a knot, in which one takes note of the strand that goes over and the strand that goes under at each crossing is a knot projection.

We can also define a knot as an embedding of S^1 into $\mathbb{E}^3$ (or S^3), which allows a further characterisation: when the embedding is differentiable, the image is called a *tame knot*. When it is not differentiable, it is said a *wild knot*.

As all knots are homeomorphic to the circle, they are all topologically equivalent as one-dimensional spaces. Knot theory can be at times more concerned with how the host space $\mathbb{E}^3$ encompasses such deformed circles than with those one-dimensional spaces themselves. Isotopy gives the complete equivalence of knots, however, it is not very practical. Given two knots, it is in general very difficult to find whether or not some isotopy exists taking one of the knots into the other. For this reason, people looked after other methods to find knots from each other.

The first characterisation of knots, weaker than the concept of

isotopy but still very powerful, is given by certain groups, which
at least have the advantage of being in principle computable. To
the effect of visualisation, it is convenient to figure the knot K
by its tubular neighborhood K, as pictured in Figs. 20 and 21.

Figure 20: The trefoil and the torus knot with a tubular
neighborhood.

The intuitive idea of twisting a real rope knot can be recovered with the following notion: K' is a *framed knot* if it has a
continuous normal vector field. This corresponds to embedding
closed twisted bands instead of circles into $\mathbb{E}^3$. *Framed links* are
introduced as an immediate generalisation.

Given a knot, K (or its tubular neighborhood K'), consider
the complement $\mathbb{E}^3\backslash K$. The *knot group* associated with K is the
fundamental group of this complement, denoted by $\pi_1(\mathbb{E}^3\backslash K)$.
Hence, the group of the trivial knot is $\mathbb{Z}$. Straightforward experiments with a rope convince the reader that such groups can
be very complicated. When two knots have the same isotopy
type, they have the same group. But two knots may share the
same knot group while being distinct, i. e., with no isotopy
equivalence between them. Less strict, but by far more practical, characterisation of link-type is given by invariants. These

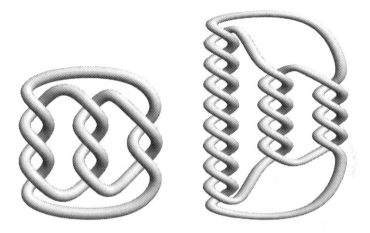

Figure 21: Pretzel knots with a tubular neighborhood, with different number of crossings and links.

can be simple numbers or, more commonly, polynomials which remain the same under an isotopy.

⇨ **Comment 2.8.** ☞ For a knot K, its mirror image is obtained by reflecting it with respect to a plane in $\mathbb{R}^3$. On a diagram, this amounts to changing all crossings to the opposite ones. For instance,

The reverse of an oriented knot is the same knot with opposite orientation. One can prove that the trefoil is not equal to its mirror image. ✓

2.2.1 Braids and links

Given a braid, one can obtain its *closure* simply by identifying corresponding initial and endpoints. Again, experiments with pieces of rope are helpful. The relation between links and braids is given by the Alexander theorem, which requires a preliminary notion. For instance, with the sole generator σ_1 of the 2-strand group B_2 one can build the Hopf link and the trefoil. They are, respectively, the braids σ_1^2 and σ_1^3 when their corresponding ends meet. Alexander theorem asserts that

☞   *Every knot or link is isotopic to a closed braid.*

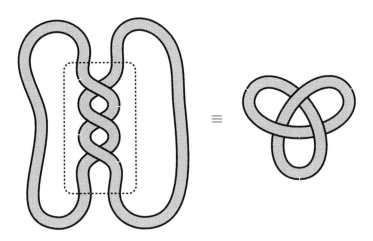

Figure 22: The trefoil as the closure of a braid.

Given the braid b which closure, denoted by $\hat{b}$, it corresponds to a link K, and one can write $\hat{b} = K$. This means that a link-type can be represented by a word with the generators of some braid group B_n. Experiments show that this correspondence is not one-to-one. Many braids can correspond to a given link-type, and the relation between braids to knots and links can be then established [Rol03, Bir75].

The relationship with braids allows us to apply to links the procedure based on Reidemeister moves to examine their eventual equivalence. This is the simplest method to study the isotopy of links. A link defines an isotopy class with all the other links obtainable from it, by a finite succession of moves RI, RII, and/or RIII Reidemeister moves. Thus, also links are ultimately studied through their diagrams and projections onto a plane. A full three-dimensional characterisation has been found by Witten, using quantum field theory [YG89]. Two remarks are in place here. The first one concerns usual links: the Reidemeister moves RI and RII simplify the links as a rule, but there exist links that admit no simplification at all. They are suitably called *demons*. The second one concerns framed links: for them, there is no RI equivalence and one is forced to restrict oneself to equivalence under moves RII and RIII. This equivalence is called a *regular isotopy* and, to distinguish from this case, a complete isotopy is sometimes called *ambient isotopy*. There exists a very important result in establishing a relationship between those distinct braids, whose closure defines the same link-type. It is the Markov theorem, stated below.

Notice that some link-type invariants come out naturally from the relationship between links and braids, and the relation of braid groups to symmetric groups. First, given a link K, one chooses a braid $b \in B_n$, whose closure leads to K, in the sense that $\hat{b} = K$. The homotopy (1.21) takes the braid b into a permutation $P = h(b)$, of some fixed cycle-type fixed by the set $\{\nu_k\}$, where ν_k denotes the number of k-cycles. The total number of cycles, $m = \nu_1 + \nu_2 + \nu_3 + \cdots + \nu_n$, is an invariant: an integer number that cannot be changed by the isotopies, which are continuous deformations. This is a rough characterisation of K, shared by many other links. A finer invariant is the set of numbers $\{\nu_k\}$ itself. Another invariant is the *writhe* $w(L)$ associated with the link L. Let us start by first defining the *sign*

$\varepsilon(c)$ of a crossing c as

$$\varepsilon\left(\underset{}{\bigtimes} \right) = 1 \quad \text{and} \quad \varepsilon\left(\underset{}{\bigtimes} \right) = -1,$$

changing signs for each further inversion of sense. Given two curves γ and σ, their *linking number* (or algebraic crossing number) — $\mathrm{lk}(\gamma, \sigma)$ — is half the sum of their crossing signs:

$$\mathrm{lk}(\gamma, \sigma) = \frac{1}{2} \sum_{c_i} \varepsilon(c_i). \tag{2.30}$$

This is an ambient isotopic invariant, of special interest to descry orientation. For example, there are two distinct Hopf links with opposite linking numbers, obtained by reversing the orientation of one of the rings while keeping the other fixed. Now, given a knot or link, its *writhe* (or *twist number*) is the sum of the signs of all its self-crossings,

$$w(L) = \Sigma_{c_i} \varepsilon(c_i). \tag{2.31}$$

Given a braid $b = \sigma_{i_1}^{a_1} \sigma_{i_2}^{a_2} \cdots \sigma_{i_p}^{a_p} \in B_n$, the writhe of a braid,

$$w(L) = \sum_{k=1}^{p} a_k, \tag{2.32}$$

is the sum of the exponents of the generators in the braid. The writhe is a regular isotopy invariant, but it is not RI invariant. Once the link is represented by a word formed with braid generators (and their inverses), the writhe is also the algebraic sum of all exponents. It is anyhow evident that all such invariants are solely partial characterisations of links, and much information is lost in the process of arriving at them.

Let us say something beyond, regarding orientation. One can define an orientation for a link K, by attributing a sense to the way it is drawn. More precisely, as braids have been defined as oriented objects, closed braids correspond to oriented links. One can describe the link by the word in the generators which defines the related closed braid. Reversal of orientation hence corresponds to the same word, even though read in the inverse sense. One can also think of changing the orientation of the host space ($\mathbb{E}^3$ or S^3), corresponding to replace each generator in the word by its inverse. Consider the union B of all the braid groups, $B = \cup_{n=1}^{\infty} B_n$. Now suppose that two braids $b_r \in B_r$ and $b_s \in B_s$ correspond to the same link-type K. Then *Markov theorem* asserts that

(a) b_r and b_s are similar: there exists some $U \in B$ such that $b_s = U b_r U^{-1}$;

(b) there exists a sequence of mappings

$$g^{\pm} : B_r \quad \rightarrow \quad B_{r+1}$$
$$b_r \quad \mapsto \quad g^{\pm}(b_r) = b_r \sigma_r^{\pm}, \qquad (2.33)$$

relating b_r and b_s.

The theorem provides link-invariants in the following way. Consider a mapping f from the braid B into some ring R (for example, the ring of integer numbers or the ring of polynomials), $f : B \rightarrow R$. Suppose that f is invariant under similarity in each B_r, and that it satisfies $f(b_r) = f(g^{+}(b_r)) = f(g^{-}(b_r))$. In this case f is called a *Markov trace*. A Markov trace is a link-type invariant. Although it is difficult in general to find whether a given invariant, found by some independent means, has this type or not, most of the known invariants have been shown to be Markov traces.

Given the braid group B_n, equivalent braids expressing the same link can be mutually transformed by successive applications of two types of operations, type I and type II Markov moves, respectively given by [Mar35, AW88]

1. $AB \mapsto BA, \qquad A, B \in B_n,$

2. $A \mapsto A\sigma_n, \quad A \mapsto A\sigma_n^{-1}, \quad A \in B_n, \sigma_n \in B_{n+1}.$

Therefore, a topological invariant link polynomial can be obtained, when one constructs a suitable representation C_n of the braid group B_n and then find a Markov move invariant quantity defined on C_n. One denotes the representation of the generators $\sigma_i \in B_n$ by $g_i \in C_n$ and a link polynomial by $\alpha(\,\cdot\,)$. The link polynomial $\alpha(\,\cdot\,)$ must satisfy the conditions:

1. $\alpha(AB) = \alpha(BA), \qquad A, B \in C_n,$

2. $\alpha(Ag_n) = \alpha(Ag_n^{-1}) = \alpha(A),$

3. $A \mapsto A\sigma_n^{-1}, \quad A \in C_n, g_n \in C_{n+1}.$

Given a braid $A \in B_n$, the operation $A \mapsto A\sigma_n^{\pm 1}$ is called stabilizing the braid and is referred to as the stabilisation move. It is worth to emphasise that stabilisation changes the number of strings in the braid by one, and one refers to both the addition and removal of a string as stabilizing. Starting with the identity braid $1 \in B_1$, it can be stabilised to yield $\sigma_1 \in B_2$. Hence, one can move between two braids in different braid groups. Markov theorem states that conjugacy and stabilisation suffice for knot equivalence. In this way, Markov theorem can be restated: given two braids $A \in B_r$ and $B \in B_s$, these braids are equivalent if and only if B can be obtained from A by a finite sequence of conjugacy or stabilisation moves. In this case, A and B are Markov equivalent braids [Mar35, Bir74, Ban09].

The link polynomial can be derived by finding the Markov trace $f(\,\cdot\,)$ on C_n, having the Markov properties [Jon85, TL71]

$$f(AB) \;=\; f(BA), \qquad A, B \in C_n, \qquad (2.34)$$
$$f(Ag_n) \;=\; \tau f(A), \qquad g_n \in C_{n+1}, \qquad (2.35)$$
$$f(Ag_n^{-1}) \;=\; \bar{\tau}(A), \qquad\qquad (2.36)$$

where

$$\tau = f(g_i), \qquad \tau = f\left(g_i^{-1}\right), \qquad i \in \{1, \ldots, n\}. \qquad (2.37)$$

Exercises

(1) Show that the writhe is invariant under RII and RIII Reidemeister moves.

(2) Show that the Euler characteristic of the sphere equals two.

(3) Prove that there are no knots with crossing number one or two.

(4) Prove that the only knots with crossing number three are the trefoil and its mirror image.

(5) Let J and K be knots. The connected sum of J and K, denoted $J\#K$, can be defined by cutting each of the knots and joining the loose edges pairwise so that no new crossings are introduced. If a knot is the composition of two knots then is called a composite knot. If a knot cannot be drawn as a connected sum of two knots, it is called a prime knot. Prove that the connected sum $J\#K$ is well-defined.

(6) Show that for any knot K, the composition $K\#\bigcirc$ of K with the unknot $\bigcirc$ is isotopic to K.

(7) Show that any two embeddings $g_1, g_2 : S^1 \hookrightarrow \mathbb{R}^3$ are homotopic [RAK15].

(8) Show that the link polynomial, defined in the last paragraph of this chapter, is related to the Markov trace by

$$\alpha(A) = (\tau\bar{\tau})^{-\frac{n-1}{2}} \left(\frac{\bar{\tau}}{\tau}\right)^{\frac{k(A)}{2}} f(A), \quad A \in C_n, \tag{2.38}$$

where $k(A)$ is the exponent sum of the elements g_i that compose the braid representation A.

(9) Show that the crossing number, c, of a $T_{p,q}$ torus knot, with $p, q > 0$, is given by

$$c = \min((p-1)q, (q-1)p). \tag{2.39}$$

Chapter 3

Polynomial invariants

As stated in Chapter 2, another characterisation of the link-type, weaker than that one given by groups, is provided by polynomial invariants. Highly commendable collections of papers on the subject are Refs. [YG89, Koh90]. The idea of somehow fixing invariance properties through polynomials is an old one. Already Poincaré used polynomials, nowadays named after him, as a shorthand to describe cohomological properties of Lie groups. For a group G, such polynomials are given by

$$p_G(t) = \beta_0 + \beta_1 t + \beta_2 t^2 + \cdots + \beta_n t^n, \qquad (3.1)$$

where $\{\beta_k\}$ is the set of Betti numbers associated with G. They are, clearly, invariant under homeomorphisms. Notice that each

β_k is the dimension of the harmonic k-forms on G, namely, the spaces of cohomology equivalence classes. The important point regarding polynomial invariants is that the coefficients evince the number of independent equivalence classes.

⇨ **Comment 3.1.** ☞ There is a prominent well known example concerning several variables invariant polynomial in physics. We have stated in Section 1.2 that the monomials $t_1^{p_1} t_2^{p_2} t_3^{p_3} \ldots t_r^{p_r}$, corresponding to equivalence classes of the symmetric group, are invariants of the symmetric group S_n. Permutations belonging to the same cycle type, i. e., with the same set $\{p_j\}$, constitute equivalence classes, as they are led to each other under the action of any element of in S_n. The total number of permutations with such a fixed cycle configuration, consisting of the number of elements in the corresponding conjugate class, reads

$$C_n = \frac{n!}{\prod_{j=1}^{n} p_j! \, j^{p_j}}. \tag{3.2}$$

The n-variable generating function for these numbers is the cycle indicator polynomial [Com74]

$$C_n(t_1, t_2, \ldots, t_n) = \sum_{i=1}^{n} \sum_{p_i} \frac{n!}{\prod_{j=1}^{n} p_j! j^{p_j}} \, t_1^{p_1} t_2^{p_2} \ldots t_r^{p_r}, \tag{3.3}$$

where the second summation is a memento of two conditions: the summation is over all values in the set $\{p_i\}$ satisfying the conditions $\sum_{j=1}^{n} p_j = m$ and $\sum_{j=1}^{n} j p_j = n$. Of course, such a summation of invariant objects is itself invariant. The coefficients of the Poincaré polynomials give the number of equivalence classes, whereas here they measure the membership of each conjugate class. ✓

⇨ **Comment 3.2.** ☞ In physics the canonical partition function of a real non-relativistic gas of N particles, contained in a d-dimensional volume V, is just such a cycle indicator polynomial. If $\beta = 1/kT$, λ is the mean thermal wavelength and d_j denotes the jth cluster integral [Pat72], that partition function reads

$$Q_N(\beta, V) = \frac{1}{N!} C_N(t_1, t_2, \ldots, t_N), \qquad t_k = k \, d_k \frac{V}{\lambda^d}. \tag{3.4}$$

The canonical partition function is thus an polynomial invariant of the symmetric group. Of course, here physics adds meaning to the dummy variables t_k. ✓ ✓

In the previous two examples, we see polynomial invariants as generating functions of monomials, one for each class. Suppose some set of objects, separated somehow into classes. The idea is to attribute a monomial t^k to the kth class of objects, multiply it by the number of elements of that class, and then sum all the results. The coefficients of the Poincaré polynomials give the number of equivalence classes, whereas in the canonical partition function they measure the membership of each class. But let us not lose the focus on knots and links.

We pass now to the ideas leading to invariant polynomials for links, as partial characterisations in the sense that two knots with coincident polynomials can be distinct or equivalent, but two links with different polynomials are surely different. The successive cases given below are progressively finer. They are sometimes able to distinguish between links that other polynomials were unable to separate.

3.1 The Alexander polynomial

The first polynomial invariants found are the Alexander polynomials. By Alexander's theorem, every oriented tame link can be obtained from some braid by closing it, that is, by uniting the respective ends. To a braid b, there corresponds a link, consisting of its closure $\hat{b}$. In principle, one should specify the braid group so that a link is denoted by the pair (b, n). The trivial link with n components,

$$\bigcirc\bigcirc\bigcirc\bigcirc\ldots\bigcirc, \tag{3.5}$$

is denoted by (I, n). The unknot comes out as

$$\bigcirc = (\sigma_1\sigma_2\sigma_3\ldots\sigma_{n-1}, n), \tag{3.6}$$

for any $n \in \mathbb{N}$. This last example illustrates, by the way, the many-to-one character of this relationship.

Given a link $\hat{b}$, with the braid b represented by a word in B_n whose sum of the exponents is the writhe $w(b)$ taken for b its matrix representative in the reduced Burau representation. Then the Alexander polynomial is given by

$$\Delta_{\hat{b}}(t) = \frac{t^{n-1-w(b)} \det[I - b]}{1 + t + t^2 + t^3 + \cdots + t^{n-1}}. \qquad (3.7)$$

Consider the simplest example. In B_2 there is only the generator σ_1, which in the reduced Burau representation is simply the one-entry matrix $\sigma_1 = [-t]$. Experience with two strings shows that:

(i) $\hat{\sigma}_1$ itself corresponds to the unknot.

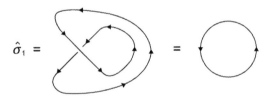

Figure 23: The unknot.

(ii) $\hat{\sigma}_1^2$ corresponds to the Hopf link (two knotted rings, linking number equals -1).

(iii) $\hat{\sigma}_1^3$ corresponds to the trefoil knot (of writhe equals 3):

But here the reduced Burau representation is one-dimensional, $\sigma_1 = [-t]$, $\sigma_1^2 = [t^2]$ and $\sigma_1^3 = [-t^3]$. We can calculate the

$$\hat{\sigma}_1^2 =$$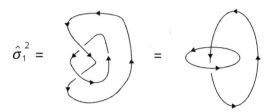

Figure 24: The Hopf link.

$$\hat{\sigma}_1^3 =$$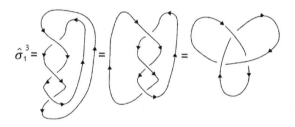

Figure 25: The trefoil knot.

polynomials trivially, respectively by:

$$
\begin{aligned}
\Delta_{\mathrm{o}}(t) &= \frac{t^{(2-1-1)/2}[1+t]}{1+t} = 1\,, \\
\Delta_{\mathrm{Hopf}}(t) &= \frac{t^{(2-1-2)/2}[1-t^2]}{1+t} = \frac{1}{\sqrt{t}} - \sqrt{t}\,, \qquad (3.8) \\
\Delta_{\mathrm{trefoil}}(t) &= \frac{t^{(2-1-3)/2}[1-t^3]}{1+t} = -t+1-\frac{1}{t}\,.
\end{aligned}
$$

This was the most practical way to calculate the Alexander polynomial up to around 1970. Then Conway found an inductive way to obtain the polynomial of a given link from those of simpler ones. Suppose three links that only differ from each other at one crossing. Call them L_+, L_- and L_o according to the crossing, as in Fig. 26.

$$\times \quad \times \quad)\,(\qquad\qquad (3.9)$$

Figure 26: The positive (L_+) and negative (L_-) crossings, and the uncrossing (L_o).

Their Alexander polynomials are hence related by

$$\Delta_{L_+}(t) - \Delta_{L_-}(t) + \frac{1-t}{\sqrt{t}}\,\Delta_{L_o}(t) = 0. \qquad (3.10)$$

This is called a *skein relation*, whose use allows polynomials for intricate links to be progressively obtained from simpler cases.

We can already here introduce a useful notation due to Kauffman, which will be extensively used later. A polynomial is indicated by a bracket $\langle\;\rangle$, wherein one draws merely that elementary link piece which is different in the concerned links. Hence, the skein relation (3.10) is written as

$$\left\langle \times \right\rangle_A - \left\langle \times \right\rangle_A + \frac{t-1}{\sqrt{t}} \left\langle \,)\,(\, \right\rangle \left\langle \,)\,(\, \right\rangle_A \qquad (3.11)$$

The subindex A indicates "Alexander". There are currently a few different families of polynomials. To the oldest family belong the Alexander polynomials, found in the 1930s [Ale28]. The way to their computation was rather involved before a skein relation was found for them. The first step in computing the Alexander polynomial of a knot is a small modification of the first step in computing the determinant. We orient the knot, label its arcs $x_1, x_2, \ldots, x_n$, like in Fig. 27, and construct an $n \times n$ matrix. Each row of the matrix, instead of containing $-1, -1, 2$ (and zeros), will contain $-1, t, 1 - t$ (and zeros).

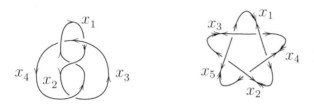

Figure 27: The 4-knot and the cinquefoil, and the orientations of their arcs.

Alexander showed that the Alexander polynomial satisfies the skein relation (3.10). Conway later demonstrated that the skein relation and a choice of value on the unknot sufficed to determine the polynomial. Conway's version is a polynomial in the z variable with integer coefficients, denoted by $\nabla(z)$, and called the Conway-Alexander polynomial.

Consider an oriented link diagram, where L_+, L_-, and L_o are link diagrams as in Fig. 26. Therefore the Conway skein relations reads

$$\nabla(L_o) = 1 \tag{3.12}$$
$$\nabla(L_+) - \nabla(L_-) = -z\nabla(L_o). \tag{3.13}$$

The relationship to the standard Alexander polynomial is given by

$$\Delta_L(t^2) = \nabla_L\left(t - t^{-1}\right). \tag{3.14}$$

Here Δ_L must be normalised by multiplication of $\pm t^{n/2}$ to satisfy the skein relation

$$\Delta(L_+) - \Delta(L_-) = \left(t^{1/2} - t^{-1/2}\right)\Delta(L_0), \qquad (3.15)$$

yielding thus a Laurent polynomial in the $\sqrt{t}$ variable.

Besides, let "+" be the diagram on the left in Fig. 29. It is a diagram of an unknot, with one (right-hand) crossing. Then "−", the diagram on the middle in Fig. 29, is also an unknot and s is an unlink, consisting of two unlinked circles. Conway formula yields

$$\nabla_+ - \nabla_- = z\nabla_s, \qquad (3.16)$$

whence $\nabla_s = 0$. Indeed, $\nabla_L = 0$ for any split link L, namely any link that splits into $n \geq 2$ pieces. Considering still the + be the diagram on the left in Fig. 29, the rectangle is not part of the knot, it is just there to show you which crossing we are looking at. Firstly, one must distinguish between right-handed and left-handed crossings, as shown in Fig. 28. Traveling on an overcross in the direction given by the orientation of the knot, if the undercross goes from right to left it is a right-handed crossing; if left to right it is a left-handed crossing.

Figure 28: Positive and negative crossing orientations.

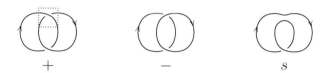

Figure 29: Links with both the orientations, and the trivial link.

Some informations are useful, and can be illustrated by Fig. 30.

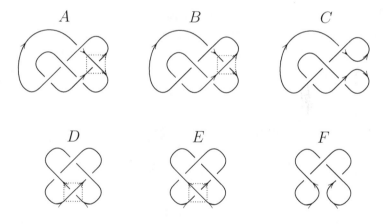

Figure 30: Links with right-handed and left-handed crossings, and uncrossing.

The links A, B, and C differ at only one crossing, where A has a right-handed crossing, B possesses a left-handed crossing, and C is the uncrossing. Hence,

$$\nabla_A - \nabla_B = -z \, \nabla_C. \tag{3.17}$$

The link B is a trefoil. Hence

$$\nabla_B = z^2 + 1. \tag{3.18}$$

On the other hand, C, as a simple right link implies that

$$\nabla_C = -z. \tag{3.19}$$

Similarly, the links D, E, and F differ at only one crossing, where D has a right-handed crossing, E possesses a left-handed crossing, whereas F corresponds to the uncrossing. Hence,

$$\nabla_D - \nabla_E = -z \, \nabla_F. \tag{3.20}$$

D is a simple right link, therefore

$$\nabla_D = -z, \tag{3.21}$$

and F is the unknot, implying that

$$\nabla_F = 1. \tag{3.22}$$

Putting it all together yields

$$\begin{aligned} \nabla_D &= 4z^2 + 1, \\ &= 4t^2 - 7t + 4, \end{aligned} \tag{3.23}$$

upon substituting $z = \sqrt{t} - \frac{1}{\sqrt{t}}$ and normalizing it.

3.1.1 More about skein relations

As asserted, after Conway's skein relations, it has become easier
to compute some Alexander polynomials. These relations, which
are different for each family of polynomials, provide an inductive way to obtain the polynomial of a given link from those
of simpler ones. Suppose three links that only differ from each
other at one crossing. We already know that there are three
types of crossing: $\diagup\!\!\!\!\diagdown$, its inverse $\diagdown\!\!\!\!\diagup$, and the identity, or
uncrossing $)\,($.

The polynomial associated with a knot K is indicated by a
bracket $\langle K \rangle$. If a knot K' differs from K only in one crossing,
then their polynomials differ by the polynomial of a third knot,
in which the crossing is abolished. There are numerical factors
in the relation, written in terms of the variable of the polynomial. Instead of drawing the entire knot inside the bracket, only
that crossing which is different is indicated. For instance, the
Alexander polynomials associated with K and K' are related by
relation (3.11). It equivalently states that the σ_j generators also
form a Hecke algebra. It is a graphic version of the equation

$$\sigma_j^{-1} - \sigma_j + \frac{t-1}{\sqrt{t}}\, I = 0. \tag{3.24}$$

Given two links L and H, the skein relation must be supple-
mented by a general rule

$$\langle HL \rangle = \langle H \rangle \langle L \rangle, \tag{3.25}$$

if H and L are unconnected parts of HL, and by normalising of
the bubble (the polynomial of the unknot), which is different for
each family of polynomials. For the Alexander polynomial,

$$\langle \bigcirc \rangle = 1. \tag{3.26}$$

A skein relation relates polynomials of different links. However, it is not in general enough for a full computation. It will be useful for the knowledge of $\langle K \rangle$ only if $\langle K' \rangle$ is better known. Kauffman extended the previous weaving patterns by introducing the so-called monoid diagrams, including objects like ⌣⌢ .

Therefore, if one adds convenient relations, like

$$\left\langle \times \right\rangle = \sqrt{t} \left\langle \; \right\rangle \; \left\langle \; \right\rangle - \left(\sqrt{t} + \frac{1}{\sqrt{t}}\right) \left\langle \; \smile \atop \frown \; \right\rangle, \quad (3.27)$$

$$\left\langle \times \right\rangle = \frac{1}{\sqrt{t}} \left\langle \; \right\rangle \; \left\langle \; \right\rangle - \left(\sqrt{t} + \frac{1}{\sqrt{t}}\right) \left\langle \; \smile \atop \frown \; \right\rangle, \quad (3.28)$$

one can go deeper and down to simpler links, and at the end only the identity and simple blobs, $\bigcirc$, remain.

The symbol ⌣⌢ represents a projector. Kauffman decomposition is justified by Jones' discovery of braid group representations in some special von Neumann algebras, which are generated by projectors. Jones has thereby found other polynomials, and also clarified the meaning of the skein relations. The cycle indicator polynomial appears as the partition function of a system of identical particles. Jones polynomials appear as the partition function of a lattice model.

3.2 The Jones polynomial

For almost half a century Alexander's were the only known polynomials. Only in the early 1980s did a new kind turn up, and that coming from a rather unexpected algebraic direction [Jon83]. In his work dedicated to the classification of von Neumann factors, Jones was led to examine certain finite dimensional von Neumann algebra A_{n+1}, generated by the identity I and by a set $\{p_1, p_2, \ldots, p_n\}$ of n projectors satisfying [Jon85, Jon87]

$$p_i^2 = p_i = p_i^\dagger, \tag{3.29}$$

$$p_i\,p_{i\pm1}\,p_i = \tau p_i, \tag{3.30}$$

$$p_i p_j = p_j p_i, \qquad \text{for } |i - j| \geq 2. \tag{3.31}$$

The parameter $\tau \in \mathbb{C}$ is called the Jones index. Jones himself wrote $\tau = \frac{t}{(1+t)^2}$, where t is another complex number, more convenient for some purposes. For more involved von Neumann algebras, the Jones index is some kind of dimension of subalgebras, in terms of which the whole algebra can be decomposed in some sense. In lattice models of statistical mechanics, with a spin variable at each vertex, the Jones index is the dimension of the spin-space. A_n is a complex algebra, with product operation (3.29 – 3.31) and an addition operation, which will be used later. Conditions (3.30, 3.31) involve clearly a nearest neighbor prescription, and are reminiscent of the braid relations. Some linear combinations of the projectors and the identity provide braid group generators.

A finite-dimensional von Neumann algebra is just a product of matrix algebras and can be represented in the direct-product notation. Consider the sequence of algebras A_n. One can add the algebra $A_0 = \mathbb{C}$ to the sequence. If one imposes that each algebra A_n embeds naturally in A_n, it turns out that this is possible for arbitrary n only if t takes on some special values: either

(a) t is real positive, or
(b) $t = e^{\pm 2\pi i/k}$, with $k = 3, 4, 5, \ldots$, implying that

$$\tau^{-1} = 4\cos^2(\pi/k). \tag{3.32}$$

For these values of t, there exists a trace defined on the union of the A_n algebras, defined as a function tr: $\bigcup_n A_n \to \mathbb{C}$, entirely determined by the conditions

$$\text{tr}(ab) \ = \ \text{tr}(ba), \qquad \forall\, a, b \in A_n, \qquad (3.33)$$

$$\text{tr}(w\, p_{n+1}) \ = \ \tau\, \text{tr}(w), \qquad \forall\, w \in A_n, \qquad (3.34)$$

$$\text{tr}(a^\dagger a) \ > \ 0, \qquad\qquad \text{if } a \neq 0, \qquad (3.35)$$

$$\text{tr}(I) \ = \ 1. \qquad\qquad\qquad\qquad\qquad (3.36)$$

Conditions (3.33 – 3.36) determine the algebra A_n up to isomorphisms.

⇨ **Comment 3.3.** ☞ A fascinating thing about the A_n algebras is that they lead to a family of polynomial invariants for knots, the Jones polynomials. But to physicists, perhaps the main point is that the partition function of the Potts model is a Jones polynomial, for a certain choice of the above variable t. This entails a relationship between lattice models and knots. ✓

A representation of the braid group B_n in the A_n algebra can be then derived. To each generator $\sigma_i \in B_n$, there is a mapping $r : B_n \to A_n$, that makes it to correspond to a member of the algebra, as

$$r(\sigma_i) = G_i = \sqrt{t}\, [tp_i - (I - p_i)]. \qquad (3.37)$$

The operators G_i are just the invertible elements of the algebra, so that the braid group B_n appears here as the group associated with the algebra A_n. The inverse of the generators read

$$G_i^{-1} = \frac{1}{\sqrt{t}}\, [t^{-1}p_i - (I - p_i)]. \qquad (3.38)$$

This relation has a simple algebraic meaning in terms of Hecke algebras. Each generator $G \in A_n$ satisfies a condition of the type

$$(G - aI)(G - bI) = 0. \qquad (3.39)$$

This means that the squared generators are linear functions of the generators, yielding a Hecke algebra. One has $P = \frac{G - aI}{b - a}$

(normalised so that $P^2 = P$) as the projector onto the eigenspace associated to the operator bI, in terms of which one can write

$$G = (b - a)P + aI = a(I - P) + bP. \qquad (3.40)$$

Recall that, given an operator A with a spectrum of eigenvalues $I_1, I_2, \ldots, I_N$, the idempotent operator

$$P_1 = \frac{(A - I_2)(A - I_3) \ldots (A - I_N)}{(I_1 - I_2)(I_1 - I_3) \ldots (I_1 - I_N)} \qquad (3.41)$$

(and an analogous definition for P_i) is a projector onto the I_1 eigenstate. In the present case there is only two eigenvalues, $t\sqrt{t}$ and $(-\sqrt{t}\,)$, and P_i and $(1 - P_i)$ suffices. The projectors read

$$p_i = \frac{G_i + \sqrt{t}}{(1 + t)\sqrt{t}}, \qquad (3.42)$$

and the condition $p_i^2 = p_i$ is equivalent to $G_i^2 = \sqrt{t}(t - 1)\,G_i + t^2$, or

$$\left(G_i - t\sqrt{t}\right)\left(G_i + \sqrt{t}\right) = 0, \qquad (3.43)$$

or still

$$t\,G_i^{-1} - t^{-1}G_i + \frac{t - 1}{\sqrt{t}}\,I = 0. \qquad (3.44)$$

What we are doing thus is to realise the braid group in a Hecke algebra, as with the Burau representation.

Let us introduce an inspiring notation, indicating each braid generator by

$$G_i = \,)\,(\qquad (3.45)$$

where it is implicit that the first line is the ith one, and all the unspecified lines are identity lines. The identity itself is indicated by

$$I = \underset{\smile}{\frown} \qquad (3.46)$$

and

$$G_i^{-1} = \diagup\!\!\!\!\!\diagdown \qquad\qquad (3.47)$$

Relation (3.44) can then be depicted as

$$t \left\langle \diagdown\!\!\!\!\diagup \right\rangle - t^{-1} \left\langle \diagup\!\!\!\!\diagdown \right\rangle - \frac{t-1}{\sqrt{t}} \left\langle\,\right\rangle \left\langle\,\right\rangle = 0. \quad (3.48)$$

This is a skein relation, emerging here as the representative of an algebraic relation. The representation of the braid group B_n takes place in a Hecke subalgebra of A_n, which this relation determines.

⇨ **Comment 3.4.** ☞ For the Alexander polynomial, the relation is $(G - \sqrt{t}I)(G + I/\sqrt{t}) = 0$, which shows that it is a particular case with $b = -1/a$. In the standard notation $(G - aI)(G - bI) = 0$, or $G = a(I - P) + bP$, one can write $G = \exp(X)$, with $X = (\log a)(I - P) + (\log b)P$. Hence, in terms of its logarithm, one can express

$$G = \left(a - \frac{\log a}{\log\left(\frac{b}{a}\right)}\right) I + \frac{b-a}{\log\left(\frac{b}{a}\right)} X. \qquad (3.49)$$

✓

With respect to Kauffman monoid diagrams, the projectors E_i for 4-strings are represented by ⌣̸⌢ and give to relation (3.37) the form

$$\left\langle \diagup\!\!\!\!\diagdown \right\rangle = \frac{t^{3/2} + t^{1/2}}{d} \left\langle \smile\!\!\frown \right\rangle - \sqrt{t} \left\langle\,\right\rangle \left\langle\,\right\rangle . \quad (3.50)$$

One can extend the previous weaving patterns, used for the braid group, to this algebra. Kauffman introduced the so-called monoid diagrams, in which projectors E_i are represented in Fig. 31, for a number $n = 4$ of strands.

$$E_1 \qquad E_2 \qquad E_3 \qquad\qquad (3.51)$$

Figure 31: Projectors for 4-strings Kauffman monoid diagrams.

A simple blob yields "d" as a multiplicative constant. Therefore, a bubble is normalised by

$$\bigcirc = d, \qquad\qquad (3.52)$$

and the projectors are normalised to this number.

The reason for the name *monoid diagrams* is solely to understand here. Projectors, with the unique exception of the identity, are not invertible. Adding projectors to the braid group generators yields a monoid, as the defining properties of a group do not hold anymore. The addition of projectors to the braid generators, or the passing into the group algebra, turns the matrix-diagram relationship into a very powerful technique.

The Jones polynomials are obtained as follows. Given a knot, one obtains it as the closure $\hat{b}$ of a braid b. Therefore the Jones polynomial reads

$$V_{\hat{b}}(t) = \left(-\frac{1+t}{\sqrt{t}} \right)^{n-1} \; \mathrm{tr} \, [r(b)]. \qquad\qquad (3.53)$$

Given a knot, draw it on the plane, with all the crossings well-defined. Choose a crossing and decompose it according to (3.48) and (3.50). Two new knots come out, which are simpler than the first one. The polynomial of the starting knot is equal to the sum of the polynomials of these two new knots. Do it again for each new knot. In this way, the polynomial is related to the polynomials of progressively simpler knots. In the end, only the identity and the blob remain.

Jones has shown that his polynomials are isotopic invariants. They satisfy the skein relation

$$tV_{L_+}(t) - t^{-1}V_{L_-}(t) + \left(\sqrt{t} - \frac{1}{\sqrt{t}}\right)V_{L_\mathrm{o}}(t) = 0 \qquad (3.54)$$

or, in bracket notation,

$$t\left\langle \diagup\!\!\!\!\diagdown \right\rangle - t^{-1}\left\langle \diagdown\!\!\!\!\diagup \right\rangle - \frac{t-1}{\sqrt{t}}\left\langle\ \right\rangle\ \left(\ \right) = 0, \qquad (3.55)$$

To explain why Jones polynomials are isotopic invariants is a straightforward task if one uses the Kauffman bracket formalism thoroughly, so that the demonstration is left to later consideration. The Jones polynomial can distinguish links that have the same Alexander polynomial. Perhaps the simplest example is the trefoil knot: there are two such knots, obtained from each other by inverting the three crossings. This inversion corresponds, in the Jones polynomial, to a transformation $t \mapsto t^{-1}$, which leads to a different Laurent polynomial. This means that the two trefoils are not isotopic. Of course, using the writhe one would state the same result.

The reason for the name "monoid diagrams" is simple. Projectors, with the sole exception of the identity, are not invertible. Adding projectors to the braid group generators does not satisfy group properties, but monoid ones. The addition of monoids to the braid generators, or the passing into the group algebra, turns the matrix-diagram relationship into a very powerful technique. We will come back to such diagrams later.

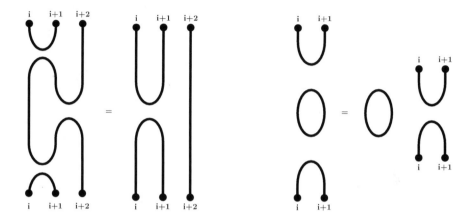

Figure 32: Conditions (3.30) and (3.29) for the projectors.

3.3 The Kauffman polynomial

The Jones procedure uses algebras somehow related to that of the braid generators to produce knot invariants. Other polynomials are obtained from another family of algebras [Jon87], known as the *Birman-Wenzl-Murakami algebras* (BWM algebras). Each algebra, denoted by $C_n(x, a)$, is a complex algebra with identity I, dependent on two complex numbers a and x. A presentation is given by the set $\{G_1, G_2, \ldots, G_{n-1}\} \subset C_n(x, a)$ of generators and their inverses, and a (non-normalised) set $\{E_1, E_2, \ldots, E_{n-1}\}$ of projectors, satisfying

$$
\begin{aligned}
G_i G_{i+1} G_i &= G_{i+1} G_i G_{i+1}, \\
G_i G_j &= G_j G_i, && \text{if } |j - i| \geq 2, \\
E_i E_{i\pm 1} E_i &= E_i, \\
E_i E_j &= E_j E_i, && \text{if } |j - i| \geq 2, \\
G_i + G_i^{-1} &= x(1 + E_i), \\
E_i G_i &= G_i E_i = a E_i \\
E_i^2 &= (a + a^{-1} - x) x^{-1} E_i =: d\, E_i, \\
E_i G_{i\pm 1} E_i &= a^{-1} E_i, \\
E_{i\pm 1} E_i G_{i\pm 1} &= E_{i\pm 1} G_i^{-1}, \\
G_{i\pm 1} G_i E_{i\pm 1} &= E_i G_{i\pm 1} G_i = E_i E_{i\pm 1} \\
G_{i\pm 1} E_i G_{i\pm 1} &= G_i^{-1} E_{i\pm 1} G_i^{-1}, \\
G_{i\pm 1} E_i E_{i\pm 1} &= G_i^{-1} E_{i\pm 1}.
\end{aligned}
\tag{3.56}
$$

The first two relations are braid relations. Making $E_i = 0$ yields a Hecke algebra. Meanwhile, the BWM algebra itself is not of this kind, as $G_i^2 - x(G_i + aE_i) + I = 0$. It is instead submitted to a cubic relation,

$$
G_i^3 - (a + x)G_i^2 - (1 + ax)G_i + aI = 0. \tag{3.57}
$$

Monoid diagrams can be introduced here. The algebra $C_n(x, a)$ embeds in $C_{n+1}(x, a)$ in the obvious way. There exists a trace, tr: $\bigcup_n C_n(x, a) \to \mathbb{C}$, uniquely defined by the four axioms:

1. tr $(I) = 1$,

2. tr $(AB) = $ tr (BA),

3. tr $(AG_n) = \frac{1}{a\,d}\mathrm{tr}(A) = \mathrm{tr}(AE_n)$,   for all $A \in C_n(x, a)$.

Consider now a link $\hat{\alpha}$ which is the closure of a braid $\alpha \in B_n$ and suppose that $\pi(\alpha)$ is the image $(\sigma_j \mapsto r(\sigma_j) := G_j)$ of α in $C_n(x, a)$. If $w(\hat{\alpha})$ is the writhe of the link $\hat{\alpha}$, then the Kauffman polynomial of this link is given by

$$F_{\hat{\alpha}}(x, a) = a^{w(\hat{\alpha})} \text{tr}[\pi(\alpha)]. \qquad (3.58)$$

At the beginning of this section, we asserted that invariant polynomials have as coefficients certain numbers giving the population of some classes. Thereafter we have sketched ways to find some such polynomials for links. Is there any topological meaning for their coefficients? Although no final and general answer is as yet available, several results found in specific models suggest the following. As already asserted, the topology involved for a knot K is that of the complement $\mathbb{E}^3 \backslash K$. The linear hole created by the extraction of K from $\mathbb{E}^3$ makes $\mathbb{E}^3 \backslash K$ a highly non-trivial multiply-connected space. Its universal covering is governed by the knot group $\pi_1(\mathbb{E}^3 \backslash K)$, other coverings being related to each one of its subgroups. The homological characteristics encoded into the Betti numbers of such coverings are isotopy invariants. At least in particular cases, the coefficients are indeed Betti numbers of some special covering of $\mathbb{E}^3 \backslash K$. It seems that, in general, given some monomial in the braid closure representation of a link, its trace yields a homology characteristic of the respective covering. Concerning the role of the algebras, Connes has shown that to any foliation corresponds a canonical von Neumann algebra, from which its topological invariants can be obtained [Con94].

3.4 The HOMFLY polynomial

One of the prominent results paved heretofore consists of whether two knot diagrams represent the same knot. Knot invariant polynomials have been seen to be computed from a knot diagram, and diagrams representing the same knot have the same polynomial. The converse may not hold, in general. Introduced in several papers sent simultaneously to the same journal, the HOMFLY polynomial is one such knot invariant and includes Alexander and Jones polynomials as particular cases. The HOMFLY polynomial is also a quantum knot invariant.

The acronym "HOMFLY" lists the authors' initials [FYH85]. They are better introduced through the main theorem:

> ☞   *There is a unique homogeneous Laurent polynomial $P_L(x, y, z)$ in the variables x, y, z, corresponding to each isotopy class L of tamed oriented links, such that*
> *a)* $xP_{L_+}(x, y, z) + yP_{L_-}(x, y, z) + zP_{L_\circ}(x, y, z) = 0$;
> *b)* *if* $P_L(x, y, z) = 1$, *then L consists of a single un-knotted component.*

The first property to be observed is that for the n-component trivial link, the HOMFLY polynomial reads

$$P_{\bigcirc\bigcirc\ldots\bigcirc}(x, y, z) = \left(-\frac{x+y}{z}\right)^{n-1}. \qquad (3.59)$$

Besides, the relationship between Alexander and HOMFLY polynomials is given by

$$\Delta_L(t) = P_L\left(1, -1, \frac{1}{\sqrt{t}} - \sqrt{t}\right), \qquad (3.60)$$

whereas the Jones polynomial is related to the HOMFLY polynomial by the expression

$$\nabla_L(t) = P_L\left(t, -t^{-1}, \frac{1}{\sqrt{t}} - \sqrt{t}\right). \qquad (3.61)$$

Since P_L is homogeneous in three variables, then it can be seen as a non-homogeneous polynomial in two variables. This can be put into many forms, the most convenient being probably by defining $P(l, m) = P_L(l, l^{-1}, m)$, in which case the skein relation becomes

$$lP_{L_+}(l, m) + l^{-1}P_{L_-}(l, m) + mP_{L_o}(l, m) = 0. \qquad (3.62)$$

Reversal of orientation in $\mathbb{R}^3$ leads to $P_L(x, y, z) \mapsto P_L(y, x, z)$ and also to $P(l, m) \mapsto P(l^{-1}, m)$. If a link L is the connected sum of links L_1 and L_2, then $P_L = P_{L_1}P_{L_2}$. A particular example is the Hopf link, whose HOMFLY polynomial reads

$$P_L = yz^{-1} + x^{-1}y^2z^{-1} - x^{-1}z. \qquad (3.63)$$

3.5 S-matrix analogy: Kauffman diagrams

The bracket polynomial, also known as the Kauffman bracket, is a polynomial invariant of framed links. Although it is not an invariant of knots or links, as it is not invariant under RI Reidemeister's moves, a suitably normalised version yields the famous knot invariant called the Jones polynomial. The bracket polynomial plays an important role in unifying the Jones polynomial with other quantum invariants.

3.5.1 Kauffman brackets

Some relationships of polynomial invariants to field theory and the S-matrix have been first suggested by Kauffman bracket model [Kau872]. In this approach, Jones polynomials come out as vacuum expectations of some operators. Given an unoriented link K, the first step is to attribute to it a 3-variable Laurent polynomial $\langle K \rangle$ recursively defined by the rules:

(1) 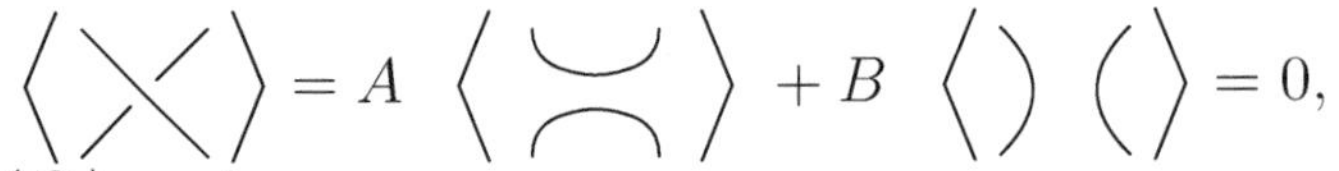 $\left\langle \diagup\hspace{-0.6em}\diagdown \right\rangle = A \left\langle \asymp \right\rangle + B \left\langle\, \right)\left(\, \right\rangle = 0,$

(2) $\langle \bigcirc \rangle = d,$

(3) $\langle \bigcirc K \rangle = d \langle K \rangle.$

The procedure is analogous to the skein relations. In these rules, only an elementary piece of an otherwise arbitrary diagram appears. Rule (1) asserts that a polynomial for a given diagram is a sum of polynomials of the same diagram, with a crossing split in the two possible trivializing ways. Rule (2) says that the value of a simple loop is d. Besides, (3) gives the relation between a diagram with a factorised loop and the remaining diagram. Given a diagram, rules (1, 2, 3) lead ultimately to an isotopy-invariant polynomial in A, B and d. One can nevertheless impose invariance under Reidemeister's moves and find constraints on the variables. With the conditions $B = A^{-1}$ and $d = -A^2 - B^2$, $\langle K \rangle$ is invariant under RII and RIII (regular isotopy). Fortunately, it is enough to normalise it conveniently to ensure also RI invariance.

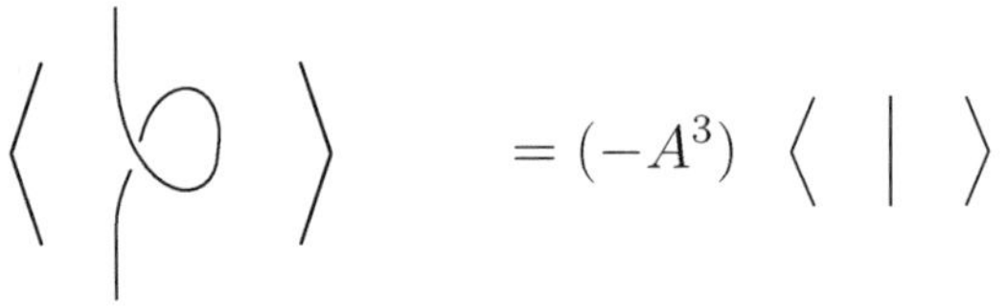

$$\left\langle \; \right\rangle \qquad = (-A^3) \left\langle \; | \; \right\rangle \qquad (3.64)$$

Figure 33: *Crossing contributes a factor* $(-A^3)$.

In fact, one can verify that each crossing, as shown in Fig. 33, contributes a factor $(-A^3)^{\pm 1}$, where the exponent is the crossing number. For any series of such crossings, one finds a factor

$(-A^3)^{w(\cdots)}$. To have also an RI invariant, one must divide by this factor for the whole link.

An ambient isotopy between two links ℓ and ℓ' is defined as the following: ℓ and ℓ' are (ambient) isotopic if there is a smooth mapping $\alpha : [0, 1] \times \mathbb{R}^3 \to \mathbb{R}^3$ such that for any value of $a \in [0, 1]$ the mapping $\alpha_a = \alpha(a, \cdot) : \mathbb{R}^3 \to \mathbb{R}^3$ is a diffeomorphism. Thus, one obtain a complete ambient isotopic invariant if one defines

$$f_K(A) = (-A^3)^{-w(K)} \langle K \rangle / \langle \bigcirc \rangle, \qquad (3.65)$$

where $w(K)$ is the writhe of K. Therefore, the Jones polynomial reads

$$V_K(t) = f_K(t^{-1/4}). \qquad (3.66)$$

This shows the isotopic invariance of $V_K(t)$.

3.5.2 The matrix-diagrammatic approach

The interpretation of the brackets as vacuum expectation values establishes a relationship with field theory in 1+1 spacetime. One can dispose a knot and consider the regions around maxima, minima, and crossing points. Each minimum is regarded as a pair creation from vacuum, each maximum consists of a pair annihilation, and each crossing corresponds to an interaction. To each one of such singularities, one attributes a matrix, with indexes spanning the values of some internal variable, as the spin, for example. One then proceeds to obtain the scattering amplitude using of the rules of a quantum mechanics: (i) concatenation is represented by matrix multiplication; (ii) if an event can happen in several disjoint alternative ways, the amplitude is the sum of the respective amplitudes.

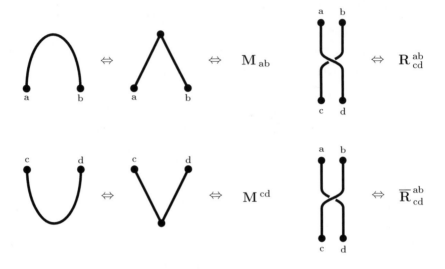

Figure 34: Indexing the singularities.

The values of the indexes, as asserted, span the possible values of some spin. Each set of values on a diagram establishes a configuration, and the sum is done over all possible configurations.

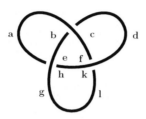

Figure 35: Indexed version of the knot previously analysed.

For the knot in Fig. 35, the final amplitude is the sum

$$T(K) = M_{ab} M_{cd} \bar{R}^{bc}_{ef} R^{ae}_{gh} \bar{R}^{fd}_{kl} M^{gl} M^{hk} \, . \qquad (3.67)$$

Given these rules, there are a few coherence requirements which restrict the matrices. They simply translate to matrices elementary properties of the diagrams. A first such constraint is given in Fig. 36,

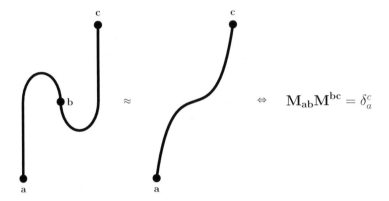

$$\Leftrightarrow \quad \mathbf{M_{ab}M^{bc}} = \delta_a^c$$

Figure 36: A first constraint on the matrices.

whereby one learns that the annihilation and the creation matrices are indeed inverse to each other. Besides, as expected from the braid relations, the interaction matrices corresponding to inverse braid generators are inverse to each other. Or, if one prefers, Fig. 37 shows a Reidemeister RII move.

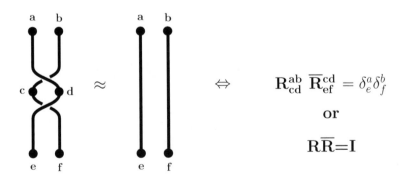

$$\Leftrightarrow \quad \mathbf{R_{cd}^{ab}} \, \overline{\mathbf{R}}_{ef}^{cd} = \delta_e^a \delta_f^b$$

or

$$\mathbf{R}\overline{\mathbf{R}}=\mathbf{I}$$

Figure 37: A second constraint: an RII move.

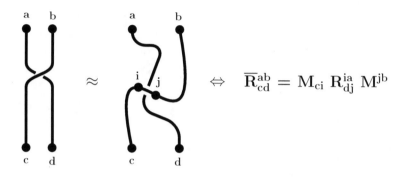

Figure 38: Braid generator and its inverse related by a consistency
requirement.

There is, of course, an analogous relation for the inverse braid
generator, as shown in Fig. 38. Finally, Fig. 39,

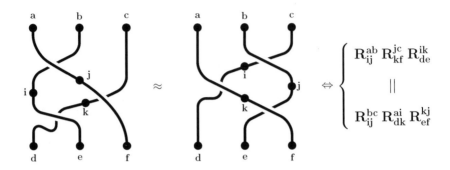

Figure 39: Yang-Baxter equation.

depicts the Yang-Baxter equation, seen here either as the braid generator basic relation or as a Reidemeister RIII move.

Taking again the relationship (1) at the first item in Subsect. 3.5.1, it is possible to write [Kau872, Kau88, Kau89, Kau91]

$$R_{cd}^{ab} = A\, M^{ab} M_{cd} + A^{-1} \delta_c^a \delta_d^b \qquad (3.68)$$

As $d = -A^2 - A^{-2}$, Fig. 36 means that the tensor $M = M^{ab} e_a \otimes e_b$ satisfies $M^2 = I$. When

$$M = \begin{pmatrix} 0 & iA \\ -iA^{-1} & 0 \end{pmatrix}, \qquad (3.69)$$

the conditions $d = -A^2 - A^{-2} = \sum_{a,b=1}^{2}(M_{ab})^2$ and $M^2 = I$ hold, and in the particular case where $M = \begin{pmatrix} 0 & i \\ -i & 0 \end{pmatrix}$, given a matrix $T = \begin{pmatrix} a & b \\ c & d \end{pmatrix}$, then

$$T \epsilon T^t = \det(T) \epsilon, \qquad \text{where } \epsilon = \begin{pmatrix} 0 & 1 \\ -1 & 0 \end{pmatrix}. \qquad (3.70)$$

When the matrix ϵ is deformed to another matrix

$$\tilde{\epsilon} = -iM = \begin{pmatrix} 0 & A \\ A^{-1} & 0 \end{pmatrix}, \tag{3.71}$$

where one keeps adopting the notation of [Kau91], it can be argued what should be the extension of the matrix T, in Eq. (3.70).

3.6 Manin's quantum plane and doorways

This section is intended to present the coalgebra underlying structure related to the concepts in Sect. 3.5.2. Some topics can be also found in Refs. [KT08, Kac02, And66, Hak02, Jag00], and references cited in this section as well.

There is a close relationship between such an extension and Manin quantum plane, consisting of a quantisation of the commutative algebra endowing a two-dimensional vector space. Let x and y coordinates in the vector space satisfying

$$xy = qyx, \tag{3.72}$$

where q is the deformation parameter, in general, a non-null complex number. More precisely, take the two-variable polynomial ring $\mathbb{C}[x, y]$ and the bilateral ideal I_q, generated by the element $xy - qyx$, where $q \in \mathbb{C}$. The quantum plane, or q-plane, is defined as being the quotient algebra

$$\mathbb{C}_q[x, y] = \mathbb{C}[x, y]/I_q. \tag{3.73}$$

One defines the q-numbers recursively by [Kac02]

$$\begin{aligned} [0]! &= 1 \\ [n]! &= \frac{q^n - 1}{q - 1}[n - 1]! \end{aligned} \tag{3.74}$$

The q-integers given by

$$[n] = \frac{q^n - 1}{q - 1} = 1 + q + q^2 + \cdots + q^{n-1} \tag{3.75}$$

are used to express an analog factorial formula,

$$[n]! = [n].[n-1].[n-2]\ldots[3].[2].[1]. \tag{3.76}$$

Given q-integers $[n]$ and $[m]$, the q-binomial is defined by the expression $\binom{n}{m}_q = \frac{[n]!}{[m]!\cdot[n-m]!}$, and consequently the Pascal q-formulas are generalised by [Kac02, KT08]

$$\binom{n+1}{m+1}_q = \binom{n}{m}_q + q^{m+1}\binom{n}{m+1}_q = q^{n-m}\binom{n}{m}_q + \binom{n}{m+1}_q,$$

for $n \geq 0$ and $1 \leq m \leq n$.

Besides, the non-commutative analog q-binomial formula

$$(x+y)^n = \sum_{j=0}^{n} \binom{n}{j}_q x^j y^{n-j} \tag{3.77}$$

is defined. In particular, the differential quantum calculus is co-variant under the action of some algebra.

⇨ **Comment 3.5.** Some extensions are realised when more generalised expressions are taken into account. While the Manin q-plane is defined intrinsically by the relations $xy = qyx$, the h-plane is defined by the relations $yx - xy = hy^2$, and the (q, h)-plane is defined by $yx = qxy + hy^2$. All these planes are precisely provided when one respectively takes the ideal I_h generated by $yx - xy - hy^2$, where $h \in \mathbb{C}$, as well as the ideal $I_{q,h}$, generated by $yx - qxy - hy^2$. The h-plane is defined as the quotient algebra $\mathbb{C}_h[x, y] = \mathbb{C}[x, y]/I_h$, and the (q, h)-plane is the quotient algebra $\mathbb{C}_{q,h}[x, y] = \mathbb{C}[x, y]/I_{q,h}$. For the variables x and y in the quantum plane $xy = qyx$, it follows that $(x+y)^n = \sum_{j=0}^{n} \binom{n}{j}_q x^j y^{n-j}$. The h-analogue is introduced as

$$(x+y)^n = \sum_{k=0}^{n} \binom{n}{k}_h y^k x^{n-k} \tag{3.78}$$

where x and y are elements in the algebra $\mathbb{C}_h[x, y]$ and the h-binomial coefficient is given by $\binom{n}{k}_h = \binom{n}{k} h^k (h^{-1})_k$, where $(a)_k = \Gamma(a+k)/\Gamma(a)$ is the shifted factorial. By historical reasons the q-shifted factorial is defined by

$$(a; q)_n = \begin{cases} 1, & n = 0, \\ \displaystyle\prod_{m=0}^{n-1}(1 - aq^m), & n \in \mathbb{N}, \end{cases}$$

and the function

$$\langle a; q \rangle_n = \begin{cases} 1, & n = 0, \\ \displaystyle\prod_{m=0}^{n-1}(1 - q^{a+m}), & n \in \mathbb{N} \end{cases}$$

is also defined [Kac02, KT08]. It can be shown that $(q^a; q)_n = \langle a; q \rangle_n$. Indeed, for $n = 0$, it follows using the definition; for $n \in \mathbb{N}$,

$$(q^a; q)_n = \prod_{m=0}^{n-1}(1 - q^a q^m) = \prod_{m=0}^{n-1}(1 - q^{a+m}) = \langle a; q \rangle_n.$$

From the l'Hôspital rule,

$$\lim_{q \to 1} \frac{(q^a; q)_n}{(1-q)^n} = \lim_{q \to 1} \frac{1-q^a}{1-q} \frac{1-q^{a+1}}{1-q} \cdots \frac{1-q^{a+n-1}}{1-q}$$
$$= a(a+1)\cdots(a+n-1). \tag{3.79}$$

Besides, it is immediate to verify that

$$\binom{n}{k}_q = \frac{\langle 1; q \rangle_n}{\langle 1; q \rangle_k \, \langle 1; q \rangle_{n-k}}, \tag{3.80}$$

for $k = 0, 1, 2, \ldots$, since

$$\binom{n}{k}_q = \frac{(1-q^n)(1-q^{n-1})\cdots(1-q^2)(1-q)}{[(1-q^k)\cdots(1-q^2)(1-q)][(1-q^{n-k})\cdots(1-q^2)(1-q)]}$$

$$= \frac{\prod_{m=0}^{n-1}(1 - q^{m+1})}{\prod_{m=0}^{k-1}(1 - q^{m+1}) \prod_{m=0}^{n-k-1}(1 - q^{m+1})} = \frac{\langle 1; q \rangle_n}{\langle 1; q \rangle_k \, \langle 1; q \rangle_{n-k}} \tag{3.81}$$

From these results, the following expression

$$\sum_{n=0}^{m}(-1)^n \binom{m}{n}_q q^{\binom{n}{2}} u^n = (u; q)_n \tag{3.82}$$

holds. Considering now the Manin quantum plane $x'y' = qy'x'$, using the linear mapping

$$\begin{pmatrix} x' \\ y' \end{pmatrix} = \begin{pmatrix} 1 & \frac{h}{q-1} \\ 0 & 1 \end{pmatrix} \begin{pmatrix} x \\ y \end{pmatrix}$$

the quantum plane can be extended by the relations

$$xy = qyx + hy^2 \qquad (3.83)$$

The quantum plane is well defined, despite the linear mapping is singular for $q = 1$. If Eq. (3.83) holds, it can be shown that [Ben99, CMP98, ZW00]:

$$
\begin{aligned}
x^k y &= \sum_{r=0}^{k} \frac{[k]_q!}{[k-r]_q!} q^{k-r} h^r y^{r+1} x^{k-r}, \\
xy^k &= q^k y^k x + h\,[k]_q\,y^{k+1},
\end{aligned}
\qquad (3.84)
$$

and also the (q, h)-binomial formula

$$(x+y)^n = \sum_{k=0}^{n} \begin{bmatrix} n \\ k \end{bmatrix}_{(q,h)} y^k x^{n-k}, \qquad (3.85)$$

where

$$\begin{bmatrix} n \\ k \end{bmatrix}_{(q,h)} = \begin{bmatrix} n \\ k \end{bmatrix}_q h^k (h^{-1})_{[k]} \qquad (3.86)$$

are the (q, h)-binomial coefficients, where $\begin{bmatrix} n \\ 0 \end{bmatrix}_{(q,h)} = 1$ and $(a)_{[k]} = \prod_{j=0}^{k-1}(a + [j]_q)$. (Remember that $[0]_q = 0$). More details can be checked in, e. g., Ref. [And66]. ✓

Now one can briefly introduce the non-commutative differential calculus in Manin plane. More details can be seen, e. g., in Ref. [Jag00]. The partial derivative operators act on the function space $\{f(x, y)\}$ that consists of polynomials in the variables x and y, expressed as $f(x, y) = \sum_{m,n} f_{mn} x^m y^n$. By taking the formal expansion

$$\frac{\partial}{\partial x} x^m = m x^{m-1}, \qquad \frac{\partial}{\partial y} y^n = n y^{n-1}, \qquad (3.87)$$

it is possible to define the differential calculus on the (x, y)-plane, where $xy = qyx$, since the commutation relations among x, y, $\frac{\partial}{\partial x}$ and $\frac{\partial}{\partial y}$ are defined:

$$\frac{\partial}{\partial x}\frac{\partial}{\partial y} = q^{-1}\frac{\partial}{\partial y}\frac{\partial}{\partial x}, \qquad \frac{\partial}{\partial x}y = qy\frac{\partial}{\partial x}, \qquad \frac{\partial}{\partial y}x = qx\frac{\partial}{\partial y},$$

$$\frac{\partial}{\partial x}x - q^2 x\frac{\partial}{\partial x} = 1 + (q^2 - 1)y\frac{\partial}{\partial y}, \qquad \frac{\partial}{\partial y}y - q^2 y\frac{\partial}{\partial y} = 1. \quad (3.88)$$

The non-commutative differential calculus on the quantum plane is covariant under the relationships [Hak02, Jag00]

$$\begin{pmatrix} x' \\ y' \end{pmatrix} = T\begin{pmatrix} x \\ y \end{pmatrix} = \begin{pmatrix} a & b \\ c & d \end{pmatrix}\begin{pmatrix} x \\ y \end{pmatrix} \qquad (3.89)$$

$$\begin{pmatrix} \frac{\partial}{\partial x'} \\ \frac{\partial}{\partial y'} \end{pmatrix} = \widetilde{T^{-1}}\begin{pmatrix} \frac{\partial}{\partial x} \\ \frac{\partial}{\partial y} \end{pmatrix} = \begin{pmatrix} d & -qc \\ -q^{-1}b & a \end{pmatrix}\begin{pmatrix} \frac{\partial}{\partial x} \\ \frac{\partial}{\partial y} \end{pmatrix} \qquad (3.90)$$

when a, b, c, d commute with x and y,

$$ab = qba, \qquad cd = qdc, \qquad ac = qca, \qquad bd = qdb,$$

$$bc = cb, \qquad ad - da = \left(q - q^{-1}\right)bc, \qquad (3.91)$$

and the equation $ad - qbc = \det_q T = 1$ holds. Note that $\det_q T$ defined in (3.91) commutes with all the elements of T, so that the requirement $\det_q T = I$ is consistent. The inverse reads

$$T^{-1} = (\det_q T)^{-1}\begin{pmatrix} d & -q^{-1}b \\ -qc & a \end{pmatrix}. \qquad (3.92)$$

The condition $ad - qbc = \det_q T = 1$, and in this case the central element $\zeta = (\det_q T)^{-1}$ in A must be included, to define the inverse of the q-matrix T_a^b (3.92), and the coinverse of the element T_a^b. The q-group is denoted by $\mathrm{GL}_q(2)$. The reader can deduce the coalgebra structure on ζ: $\Delta(\zeta) = \zeta \otimes \zeta$, $\epsilon(\zeta) =$

1, $S(\zeta) = \det_q T$. More aspects, including a quantum Clifford algebra approach and bracket deformations, can be checked in Refs. [AGR14, BR07, BR08, FRS16, RBV10]. The relations discussed can be also obtained when the conditions

$$T\tilde{\epsilon}T^t = \tilde{\epsilon}, \qquad T^t\tilde{\epsilon}T = \tilde{\epsilon} \tag{3.93}$$

are imposed, when $q = A^2$ in Eq. (3.71) [Kau91].

Exercises

(1) Is the trefoil equivalent to its mirror image? (Hint: compute the Alexander polynomial for both knots, remembering that the trefoil is the closure of σ_1^3, whereas its mirror image is the closure of σ_1^{-3}.)

(2) Compute the Jones polynomial for the 2-braid σ_1.

(3) Show that the Alexander polynomial of the torus knot $T_{p,q}$ reads

$$\Delta_{T_{p,q}} = \frac{(1 - t^{pq})(1 - t)}{(1 - t^p)(1 - t^q)}. \tag{3.94}$$

(4) Show that the Jones polynomial is a knot invariant.

(5) Prove that the Kauffman bracket is invariant under RIII Reidemeister move.

(6) Determine the Jones polynomial of the positive Hopf link, with linking number $+1$, using the Jones skein relation. Besides, show that it is not equivalent to its mirror image [RAK15].

(7) Find the Jones polynomial of the connected sum of two knots, in terms of the Jones polynomial of each of the two knots.

(8) Denoting by T the right-handed trefoil, show that

$$V_T(t) = t - t^3 - t^4. \tag{3.95}$$

(9) Do the Kauffman bracket and writhe depend on the orientation of a diagram?

(10) Show that the Jones polynomial of a knot does not depend on its orientation.

(11) Determine the HOMFLY polynomial of the trefoil.

(12) Show that the HOMFLY polynomial is a knot invariant, in the sense that it does not depend on the orientation of the knot.

(13) Show that the Jones polynomial of a right-handed torus knot $T_{p,q}$ reads

$$t^{(p-1)(q-1)/2} \frac{(1 - t^{p+1} - t^{q+1} + t^{p+q})}{1 - t^2}. \tag{3.96}$$

Chapter 4

Statistical physics connection

4.1 The Ising model

Most of the lattice models suppose a set of lattice sites, each with a set of adjacent sites, a graph, forming a d-dimensional lattice with N vertexes (sites) and some spacing (the lattice parameter) between them. In each site, a molecule is placed, endowed with some internal discrete degree of freedom, generically taken as the magnetic dipole moments of atomic spin. The lattice can be, for instance, hexagonal, or cubic centered in the faces, among the most prominently useful models. The spin at the site "k" is described by a q-dimensional vector σ_k, as illustrated in the plot on the right in Fig. 40. More precisely, a spin configuration, $\sigma = \sigma_k$, where k is a lattice site, is an assignment of a spin value to each lattice site. For any two adjacent sites i, j there is an interaction J_{ij}. Also, a site j has an external magnetic field $\mathbf{H}_j$ interacting with it. The energy of a configuration is given by the Hamiltonian function. The interaction takes place along the edges (bonds) and is given by a general Hamiltonian of the form

$$\mathcal{H} = -\sum_{i<j} J_{ij}\, \sigma_i \cdot \sigma_j - \mu \sum_k \mathbf{H}_k \cdot \sigma_k, \qquad (4.1)$$

where μ denotes the magnetic moment. The factor J_{ij} represents

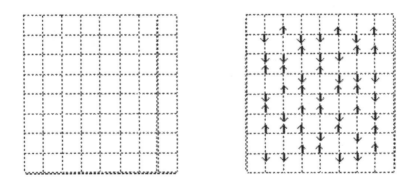

Figure 40: A spin lattice.

the coupling between the molecules situated in the ith and the jth sites. To solve such a model means to obtain the explicit form of the partition function. A reasonably realistic case is the Heisenberg model, for which $d = q = 3$. Though analytic solutions are difficult to be obtained, many results can be elicited by numerical methods. Some other cases were solved, and none of them realistic enough instead. The problem is less difficult in the nearest-neighbor approximation, which supposes that $J_{ij} \neq 0$ only when i and j are immediate neighbors. For one-dimensional systems, with no external magnetic field ($\mathbf{H} = 0$), exact solutions are known for all values of the spin dimension q. For higher dimensions, the best known model is the celebrated Lenz-Ising model, for which $q = 1$ ($s_k = \sigma_k = \pm 1$) and the same interaction is assumed for each pair of neighbors:

$$\mathcal{H} = -J \sum_{\langle ij \rangle} s_i s_j - \mathbf{H} \sum_k s_k. \qquad (4.2)$$

The symbol $\langle ij \rangle$ indicates that the summation takes place merely on the nearest neighbors. Ising models can be classified according to the sign of the interaction. For a pair i, j of indexes $J_{ij} > 0$, the interaction is called ferromagnetic, whereas if $J_{ij} < 0$, it consists of an antiferromagnetic interaction. When $J_{ij} = 0$, the spins are noninteracting.

4.2 Spontaneous symmetry breaking

The main interest in these models lies in the study of phase transitions. The one-dimensional solution

$$Q_N(\beta) = (2 \cosh \mathcal{J})^N = [2 \cosh(\beta J)]^N \qquad (4.3)$$

shows no transition at all, which is an example of a general result, known as the van Hove theorem. It asserts that one-dimensional systems with short-range interactions among constituents exhibit no phase transition. The two-dimensional Onsager solution, however, shows a transition, implemented by the behavior of the specific heat, C, whose derivative exhibits a singularity near the critical (Curie) temperature $kT_c \approx 2,269\,J$. The specific heat itself behaves, near this temperature, as $C \approx$ constant

$\times \log |T - T_c|^{-1}$. A logarithmic singularity is considered to be a weak singularity. The magnetisation, M, however, has a more abrupt behavior, of the form

$$\frac{M}{N\mu} \approx 1.2224 \left(\frac{T_c - T}{T_c}\right)^{\beta}, \quad \text{with } \beta = \frac{1}{8}. \quad (4.4)$$

Numerical studies show that in the three-dimensional case the phase transition is more accentuated. For a cubic lattice, the specific heat near the critical temperature behaves as

$$C \approx |T - T_c|^{-\alpha}, \quad \text{with } \alpha \approx 0.125, \quad (4.5)$$

Thus, general behavior depends strongly on the dimension. This transition can be thought of as a ferromagnetic transition, with an abrupt change from a state in which the magnets are randomly oriented, at high temperatures, to a microscopically anisotropic state in lower temperatures, in which the magnets are aligned altogether. It is also an order-disorder transition. The magnetisation of some real metals is qualitatively described by the Ising model. The fractional values of the exponents (4.4) tell us that these critical points are not singular points of the simple Morse type, which would have a polynomial aspect. They display degenerate points, and are obtained via the far more sophisticated procedures of the renormalisation group.

With an energy of the form $E = -J\sum_{\langle ij \rangle} s_i s_j$, there are two configurations with the (same) minimum energy: either all the spins up (+1) or all spins down (−1). Thus, at zero temperature, there are two possible states, and the minimal entropy is not zero, but $S_0 = k \log 2$. When the temperature is high, the system is in complete microscopic disorder, with its spins pointing along in all directions. Even small domains of the medium, that are large enough if compared to molecular dimensions, exhibit this isotropy, i. e., rotational symmetry. As the temperature goes down, there is a critical value at which the system chooses one of the two possible orientations and becomes spontaneously magnetised while proceeding toward the chosen fundamental state. The original macroscopic rotational symmetry of the system breaks down. Notice that the Hamiltonian is, and remains, rotationally symmetric. The word "spontaneous" acquired for this reason a more general meaning.

One calls "spontaneous symmetry breaking" every symmetry breaking which is due to the existence of more than one ground state. The fundamental state is called "vacuum" in field theory. When it is multiple, one says that the vacuum is *degenerate*. There is thus a spontaneous breakdown of symmetry whenever the vacuum is degenerate. A quantity like magnetisation, which vanishes above the critical temperature and is different from zero below it, is an order parameter. The presence of an order parameter is typical of phase transitions of the second kind, more commonly called *critical phenomena*.

4.3 The Potts model

The Potts model can be defined on any graph, that is, any set of vertexes (sites) with only (at most) one edge between each pair [Bax82]. This set of sites and edges constitutes the basic lattice, which ideally models some crystalline structure. A variable s_i, taking on N values, is defined on each site labeled by the index "i". For simplicity, again this variable is called spin. Dynamics is then introduced by supposing that only adjacent spins interact, with interaction energy $e_{ij} = -J\delta_{s_i s_j}$. The total energy reads

$$E = -J\Sigma_{(ij)}\delta_{s_i s_j}, \tag{4.6}$$

the summation on all the edges (i, j). Hence, with $\mathcal{K} \equiv J/kT$, the partition function for a M-site lattice is given by

$$Q_M = \sum_s \exp\left(\mathcal{K} \sum_{(ij)} \delta_{s_i s_j}\right), \qquad (4.7)$$

where the summation being over all the possible configurations $s = (s_1, s_2, \ldots, s_M)$. The Ising model, with cyclic boundary conditions, is the particular case with $N = 2$ and $\mathcal{K}$ replaced by $2\mathcal{K}$. Despite the great generality of this definition on generic graphs, only lattices formed with squares will be here approached. The main point in what follows is that Q_M can be obtained as the sum of all the entries of a certain transfer matrix, T. This matrix T turns out to be factorised into the product of simpler matrices, the local transfer matrices, which are intimately related to the projectors E_i of the Temperley-Lieb algebra (Comment 4.3). A surprising outcome is that the partition function for the Potts model can be obtained as a Jones polynomial for a knot related to the lattice directly.

The algebra A_n is essentially that used by Temperley and Lieb [TL71] in their demonstration of the equivalence between the ice-type and the Potts models [Bax82], notwithstanding the difference in the projector normalisation. One defines new projectors

$$E_i = \delta p_i, \qquad (4.8)$$

where δ is a numerical parameter. If $\tau = \delta^2$, the conditions become

$$\begin{aligned} E_i^2 &= \delta E_i, & (4.9) \\ E_i E_{i\pm 1} E_i &= E_i, & (4.10) \\ E_i E_j &= E_j E_i, & \text{for } |i - j| \geq 2. & (4.11) \end{aligned}$$

The Temperley-Lieb algebra A_n is formally an algebra over the ring $\mathbb{Z}[A, A^{-1}]$ of polynomials in A and A^{-1}, with integer coefficients. To understand the Temperley-Lieb algebra, consider elements in A_n being diagrams similar to braids. An element of A_n can be represented as follows by taking n points each on a left plane and a right plane in $\mathbb{R}^3$, with strings connected to them. However, unlike in the case of braids, the strings are allowed to loop backward and do not necessarily have to move from one side to the other. As usual, this is much easier to understand with a diagram.

The algebra A_n is defined as the set of all tangles satisfying the relations

$$\left\langle \times \right\rangle = A \left\langle \smile\frown \right\rangle + A^{-1} \left\langle \right\rangle \left\langle \right\rangle, \qquad (4.12)$$

and

$$K \bigcup \bigcirc = \delta K, \quad \text{where} \quad \delta = -A^2 - A^{-2}. \qquad (4.13)$$

Here K denotes an element of A_n. The generators E_i can be depicted as

$$E_1 \qquad\qquad E_2 \qquad\qquad E_{n-1}$$

We are now in a position to provide the representation of the braid group by the Termperley-Lieb algebra. One defines the homomorphism $\phi_n : B_n \to A_n$ by

$$\phi_n(\sigma_i) = AI_n + A^{-1}E_i, \qquad \phi_n(\sigma_i^{-1}) = A^{-1}I_n + AE_i, \quad (4.14)$$

for $I_n \in A_n$ representing the identity operator. The homomorphism ϕ_n can be thought of as being a representation of the Artin braid group. To realise it, one verifies that ϕ_n preserves the relations on the generators σ_i of the braid group B_n. Indeed,

$$
\begin{aligned}
\phi_n(\sigma_i)\phi_n(\sigma_i^{-1}) &= (A + A^{-1}E_i)(A^{-1} + AE_i) \\
&= I + (A^{-2} + A^2)E_i + E_i^2 \\
&= I + (A^{-2} + A^2)E_i + \delta E_i \\
&= I + (A^{-2} + A^2)E_i + (-A^{-2} - A^2)E_i \\
&= I.
\end{aligned}
$$

Besides,

$$
\begin{aligned}
\phi_n(\sigma_i\sigma_{i+1}\sigma_i) &= (A+A^{-1}E_i)(A+A^{-1}E_{i+1})(A+A^{-1}E_i) \\
&= (A^2+A^{-1}E_{i+1}A+E_i+A^{-2}E_iE_{i+1})(A+A^{-1}E_i) \\
&= A^3+AE_{i+1}+AE_i+A^{-1}E_iE_{i+1}+A^{-1}E_i^2+AE_i \\
&\quad +A^{-1}E_{i+1}E_i + A^{-3}E_iE_{i+1}E_i \\
&= A^3 + AE_{i+1} + (A^{-1}\delta + 2A)E_i \\
&\quad +A^{-1}(E_iE_{i+1} + E_{i+1}E_i) + A^{-3}E_i \\
&= A^3 + A(E_{i+1} + E_i) + A^{-1}(E_iE_{i+1} + E_{i+1}E_i) \\
&= \phi_n(\sigma_i\sigma_{i+1}\sigma_i), \tag{4.15}
\end{aligned}
$$

since it is evident that Eq. (4.15) is symmetric under transposition of i and $i+1$. Besides, if $|i-j| > 1$, it yields

$$
\begin{aligned}
\phi_n(\sigma_i\sigma_j) &= \phi_n(\sigma_i)\phi_n(\sigma_j) = (A + A^{-1}E_i)(A + A^{-1}E_j) \\
&= (A + A^{-1}E_j)(A + A^{-1}E_i) \\
&= \phi_n(\sigma_j\sigma_i). \tag{4.16}
\end{aligned}
$$

The function ϕ_n resolves the crossings of the braid, where each σ_i in the braid group represents a crossing of the strands in a braid. When one takes ϕ_n of it, the sum of the two possible resolutions of the crossing with the coefficients of A or A^{-1} is evinced, depending on which type of crossing are being resolved. Let us depict

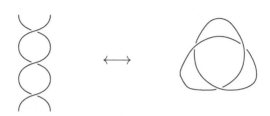

and analyze it. The trefoil is ambient isotopic to $\overline{\sigma_1^3}$. One can resolve crossings using the relation

$$
\begin{aligned}
\phi(\sigma_1^3) &= (A + A^{-1}E_i)^3 \\
&= A^3 + 3AE_1 + 3A^{-1}E_1^2 + A^{-3}E_1^3. \quad (4.17)
\end{aligned}
$$

Given a square lattice as that in Fig. 40, with the interactions just defined, consider the $N^n \times N^n$ matrices $E_1, E_2, \ldots, E_{2n-1}$, with

$$
(E_{2i-1})_{s,s'} = \frac{1}{\sqrt{N}} \prod_{j \neq i = 1}^{n} \delta_{s_j s_j'}, \quad (4.18)
$$

$$
(E_{2i})_{s,s'} = \sqrt{N} \, \delta_{s_i s_{i+1}} \prod_{j=1}^{n} \delta_{s_j s_j'}.
$$

Let us give some examples, with the notation $|s\rangle = |s_1, s_2, s_3, \ldots, s_n\rangle$:

$$
\begin{aligned}
\langle s|E_2|s'\rangle &= \delta_{s_1 s_2}(\delta_{s_1 s_1'} \delta_{s_2 s_2'} \ldots \delta_{s_n s_n'}), \\
\langle s|E_4|s'\rangle &= \delta_{s_2 s_3}(\delta_{s_1 s_1'} \delta_{s_2 s_2'} \ldots \delta_{s_n s_n'}), \\
&\vdots \qquad\qquad\qquad \vdots \\
\langle s|E_{2n-2}|s'\rangle &= \delta_{s_{n-1} s_n}(\delta_{s_1 s_1'} \delta_{s_2 s_2'} \ldots \delta_{s_n s_n'}),
\end{aligned}
$$

$$\langle s|E_1|s'\rangle = \frac{1}{N}\,\delta_{s_2 s_2'}\ldots\delta_{s_n s_n'},$$

$$\langle s|E_3|s'\rangle = \delta_{s_1 s_1'}\frac{1}{N}\,\delta_{s_3 s_3'}\delta_{s_4 s_4'}\ldots\delta_{s_n s_n'},$$

$$\langle s|E_5|s'\rangle = \delta_{s_1 s_1'}\delta_{s_2 s_2'}\frac{1}{N}\,\delta_{s_4 s_4'}\ldots\delta_{s_n s_n'},$$

$$\vdots \qquad\qquad \vdots$$

$$\langle s|E_{2n-1}|s'\rangle = \delta_{s_1 s_1'}\delta_{s_2 s_2'}\ldots\delta_{s_{n-1} s_{n-1}'}\frac{1}{N}\,. \qquad (4.19)$$

Thus, in the direct product notation, if I denotes the identity $N \times N$ matrix, the even-indexed matrices read

$$E_{2i} = \sqrt{N}\,\delta_{s_i s_{i+1}} I \otimes I \otimes I \otimes \cdots \otimes I = \sqrt{N}\,\delta_{s_i s_{i+1}}(I^{\otimes n}). \quad (4.20)$$

The matrix E_{2i} is, thus, a diagonal matrix, with entries $\sqrt{N}\,\delta_{s_i s_{i+1}}$. The odd-indexed matrices are

$$E_{2i-1} = I \otimes I \otimes \cdots \otimes \left[\frac{I}{\sqrt{N}}\right] \otimes \cdots \otimes I \otimes I$$

$$= I^{\otimes(i-1)} \otimes \left[\frac{I}{\sqrt{N}}\right] \otimes I^{\otimes(n-i)}, \qquad (4.21)$$

where $\frac{1}{\sqrt{N}}$, which is in the ith position, is a $N \times N$ matrix, also a projector, with all the entries equal to $\frac{1}{\sqrt{N}}$. The notation is purposeful: the set $\{E_k\}$ satisfies just the defining relations of the Temperley-Lieb algebra with $M = 2n - 1$, and Jones index equals N. The Jones index is in this case just the dimension of the spin space.

One introduces the local transfer matrices

$$V_j = I + \frac{v}{\sqrt{N}}\,E_{2j}, \qquad W_j = \frac{v}{\sqrt{N}}\,I + E_{2j-1}\,, \qquad (4.22)$$

and $v = e^{\mathcal{J}} - 1$. One can also introduce the Kauffman decomposition

$$\left\langle \times \right\rangle = \left\langle \,\rangle\,(\, \right\rangle + \frac{v}{\sqrt{N}} \left\langle \smile\!\!\frown \right\rangle, \qquad (4.23)$$

which means that the inverse reads

$$\left\langle \times \right\rangle = \frac{v}{\sqrt{N}} \left\langle \,\rangle\,(\, \right\rangle + \left\langle \smile\!\!\frown \right\rangle, \qquad (4.24)$$

The bubble normalisation is given by the identification

$$\bigcirc \sim \sqrt{N}. \qquad (4.25)$$

Two global transfer matrices can be written as

$$
\begin{aligned}
V &= \exp\left[\mathcal{K}(E_2 + E_4 + \cdots + E_{2n-2})\right] \\
&= \exp\left(\mathcal{K}\sum_{j=1}^{n-1}\delta_{s_j s_{j+1}}\right)(E^{\otimes n}) \\
&= \prod_{j=1}^{n-1}\left[I + \frac{v}{\sqrt{N}}E_{2j}\right] = \prod_{j=1}^{n-1}\left[\overset{2j}{)}\overset{2j+1}{(} + \frac{v}{\sqrt{N}}\overset{2j}{\underset{\cap}{\cup}}\overset{2j+1}{}\right] \\
&= \prod_{j=1}^{n-1}\left[\overset{2j}{\times}\right]
\end{aligned}
\qquad (4.26)
$$

and

$$
\begin{aligned}
W &= \prod_{j=1}^{n}\left[vI + \sqrt{N}E_{2j-1}\right] \\
&= N^{n/2}\prod_{j=1}^{n}\left[\frac{v}{\sqrt{N}}\overset{2j-1}{)}\overset{2j}{(} + \overset{2j-1}{\underset{\cap}{\cup}}\overset{2j}{}\right] \\
&= N^{n/2}\prod_{j=1}^{n}\left[\overset{2j-1}{\times}\right].
\end{aligned}
\qquad (4.27)
$$

One now looks at these transfer matrices in terms of the $(2n-1)$ generators of the braid group B_{2n}. They are

$$V = \sigma_2 \sigma_4 \ldots \sigma_{2n-2}, \qquad W = \sigma_1^{-1} \sigma_3^{-1} \ldots \sigma_{2n-1}^{-1}. \qquad (4.28)$$

In the case of a $n \times m$ Potts lattice, the partition function reads

$$Q_{nm} = \xi^T VWVW \ldots V\xi = \xi^T T\xi, \qquad (4.29)$$

where ξ is a column vector whose all entries are equal to 1. There are m elements V and $(m-1)$ elements W in the product. To sandwich the matrix T between ξ^T and ξ is a simple trick. It means that one sums all the entries of T.

Let us now try to translate all this into the diagrammatic language. The sum over all configurations is already accounted for in the matrix product, as the index values span all the possible spin values. The question which remains is: how to put into the matrix-diagrammatic language the summation over the entries of the overall transfer matrix? The solution comes from the use of projectors. To see it, let us take for instance the case $n = m = 2$. In this case, the diagrams have four strands,

$$V = \left(I + \frac{v}{\sqrt{N}} E_2\right), \qquad (4.30)$$

$$W = (vI + \sqrt{N} E_1)(vI + \sqrt{N} E_3). \qquad (4.31)$$

The matrix involved is an element of B_4, consisting of $VWV = \sigma_2 \sigma_1^{-1} \sigma_3^{-1} \sigma_2$. One has to sum over all the values of the indexes a, b, c, d, e, and f in Fig. 41, on the left.

The result wished for is obtained by adding projectors before and after the diagram as in the center diagram of Fig. 41, and then taking the trace, that is, closing the final diagram. This closure is represented by taking identical labels for the corresponding extreme points, which is just closure in the sense of knot theory.

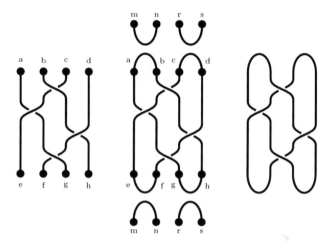

Figure 41: Matrix T with indexes (left plot), with added projectors (central plot), and its closure (right plot).

Of course, there are two extra factors from the bubbles, that must be extracted. This solution is general: for $M = 2n$ vertexes, one adds $2n$ projectors and then close the result, obtaining n extra bubbles which must then be extracted. Thus, the partition function is given by

$$Q = N^{n/2}\langle K \rangle. \tag{4.32}$$

The general relationship is thus the following: given a lattice, draw its medium alternate link K, which weaves itself around the vertexes going alternatively up and down the edges. Fig. 41 (right plot) shows the case when $m = n = 2$, which corresponds simply to T closed by pairs of cups. Each edge of the lattice corresponds to a crossing, a generator of B_n, or its inverse. Vertexes correspond to regions circumvented by loops. With the convenient choice of variables given above, the partition function is the Jones polynomial of the link.

In these lattice models, the lattice itself has been taken as fixed and regular, whose dynamics is concentrated in the interactions between the spin variables in the vertexes. In the study of elastic media and glasses, this regularity is weakened and the variables at the vertexes acquire different values and meanings.

4.4 The star-triangle relation

Given a knot drawn in some convenient two-dimensional projection, one starts by darkening alternate circumvented regions, as in Fig. 42. One then draws a graph, where the darkened regions

Figure 42: A knot, and its shadowed version.

are represented by vertexes and the crossings are given by edges. The edges are attributed signs, according to the convention given in Fig. 43.

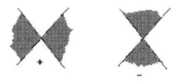

Figure 43: Conventional signs attributed to the edges.

The signs would correspond, in the statistical models, to the energy contribution of the edge to the exponential in the partition function. Now the main point: the star-triangle relation asserts that the contribution of the star and the triangle in the figure are equivalent. Their corresponding shaded drawings are shown in Fig. 44, as well as a RIII Reidemeister move, which shows that they are indeed equivalent.

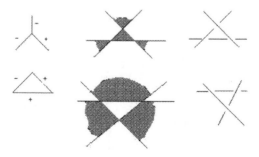

Figure 44: Contributions of the stars and the triangles are equivalent.

Thus, the star-triangle relation corresponds just to a Reidemeister RIII equivalence of links.

In Ising models, Yang-Baxter equations spell the star-triangle relation. This is more easily seen in the honeycomb lattice model, in Fig. 45, in which there appears an equivalence between adding a Δ and a Y configuration.

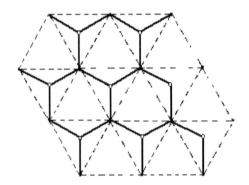

Figure 45: The honeycomb lattice model.

4.5 Cayley trees, Bethe lattices and Weyl operators

Suppose one builds up a graph in the following way: take a point p_0 as an original vertex and draw q edges starting from it. To each new extremity, add again $(q-1)$ edges. Thus q is the coordination number, or degree of each vertex in the terminology of Sect. 1.11. The first q vertexes constitute the first shell, the added $q(q-1)$ ones form the second shell. Proceeding iteratively in this way, adding $(q-1)$ edges to each point of the rth shell to obtain the $(r+1)^{\text{th}}$ shell. There are $q(q-1)^{r-1}$ vertexes in the rth shell. Suppose one stops in the nth iteration. The result is a tree with

$$V = \frac{q[(q-1)^n - 1]}{q-2}.$$ (4.33)

This graph is called a Cayley tree, as depicted in Fig. 46.

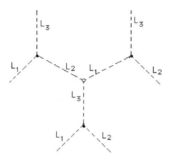

Figure 46: A tree.

More generally, a tree in which each non-leaf graph vertex has a constant number of branches n is called an n-Cayley tree. It is used in statistical mechanics, each vertex being taken as a particle endowed with spin. The partition function consists of the sum over all possible spin configurations. There is a problem, though. The number of vertexes in the nth shell is not negligible to V, so that one of the usual assumptions of the thermodynamical limit, that border effects are negligible, is jeopardised. One solution is to take $n \to \infty$, consider averages over large regions, not including last-shell vertexes, and take them as representatives of the whole system. The tree so obtained is called the Bethe lattice and the model is the Ising model itself on the Bethe lattice. Its interest is twofold: (i) it is exactly solvable and (ii) it is a first approximation to models with more realistic lattices, encompassing square, cubic, and other ones.

Formally, a Bethe lattice is an infinite connected cycle-free graph where the vertices all have the same valence. In other words, each node is connected to z neighbours, and z is called the coordination number. With one node chosen as root, all other nodes are seen to be arranged in shells around this root node, which is then also called the origin of the lattice. The number of nodes in the kth shell is given by

$$N_k = z(z-1)^{k-1}, \qquad k > 0. \tag{4.34}$$

The Bethe lattice can be thought of as being an unrooted tree, since any vertex will serve equally well as a root. The Bethe lattice where each node is joined to $2n$ others is essentially the Cayley graph of a free group on n generators. It is an infinite Cayley tree.

Now, to introduce Weyl operators, one starts by defining a circulant matrix[1] A, whose representation in some basis reads

$$\begin{pmatrix} a_N & a_{N-1} & a_{N-2} & \cdots & a_1 \\ a_1 & a_N & a_{N-1} & \cdots & a_2 \\ a_2 & a_1 & a_N & \cdots & a_3 \\ \vdots & \vdots & \vdots & \ddots & \vdots \\ a_{N-1} & a_{N-2} & a_{N-3} & \cdots & a_N \end{pmatrix}, \tag{4.35}$$

where the a_i are complex numbers such that $a_N = a_0$ and $a_{N+i} = a_i = a_{i-N}$. The matrix (4.35) has entries satisfying $A_{ij} = a_{i-j}$, which have the property

$$A_{i+k,j+k} = A_{ij}. \tag{4.36}$$

Indicating by $\{\lambda_i\}$ the eigenvalues spectrum of A, it follows that $\det(A) = \prod_{k=1}^{N} \lambda_k$. The circulant matrices eigenvalues are the

---

[1] A circulant matrix is a square matrix in which each row vector is moved away one element to the right relative to the preceding row vector.

Fourier transform of a_i in the set $\mathbb{Z}_N$:

$$\lambda_j = \sum_{k=1}^{N} a_k e^{i(2\pi/N)jk} . \qquad (4.37)$$

In particular,

$$\lambda_0 = \lambda_N = \sum_{i=1}^{N} a_i, \qquad (4.38)$$

$$a_0 = a_N = \frac{1}{N}\sum_{i=1}^{N} \lambda_k = \frac{1}{N}\text{tr } A. \qquad (4.39)$$

For more details, one can check Ref. [Ald01].

A Fourier transform is well-known to take a function $a : j \mapsto a(j) = a_j$ in a group into a function $\lambda : k \mapsto \lambda_k$ in any another group. This property is called the Pontryagin duality. A circulant matrix is fixed by the alphabet $(a_1, a_2, \ldots, a_N)$, and its eigenvalues are fixed by the alphabet $(\lambda_1, \lambda_2, \ldots, \lambda_N)$. The Fourier transform (4.37) leads an alphabet to the other, which reflects the cyclic groups isomorphic to $\mathbb{Z}_N$ are self-dual, in the Pontryagin duality context.

Given a fixed arbitrary natural number $N \in \mathbb{N}$, it is convenient to denominate $\omega = \exp(2\pi i/N)$ the primitive N^{th}-root of the unity, in such a way that

$$\lambda_k = \sum_{i=1}^{N} a_i \omega^{ki} , \qquad (4.40)$$

$$a_i = \sum_{k=1}^{N} \lambda_k \omega^{-ki}. \qquad (4.41)$$

Let U and V be two unitary operators that satisfy the relation

$$VU = \omega UV, \qquad \omega \in \mathbb{C}. \qquad (4.42)$$

Finite dimensional representations are obtained when one takes $N \times N$ matrices representing U and V in such a way that $U^N = I$ and $V^N = I$. It is immediate to realise that $\det U \neq 0 \neq \det V$, which implies that $\omega^N = 1$.

The group elements are therefore products of powers of type $\omega^p U^m V^n$, that are related to the relations that define the Heisenberg group in terms of 3-tuples

$$(m, n, p) \star (r, s, q) = \left(m + r, n + s, p + q + \frac{1}{2}(nr - ms)\right). \quad (4.43)$$

These relations hold when one identifies

$$(m, n, p) = \omega^p \omega^{mn/2} U^m V^n. \quad (4.44)$$

Such polynomials in the U and V variables constitute a complete basis for quantum operators related to some physical system. A quantum operator is an element of a group ring in some representation of the Heisenberg group. In a basis for the space state of orthonormal vectors $|v_k\rangle$, where k is an integer defining the cyclic condition $|v_k + N\rangle = |v_k\rangle$, the operator U can be defined by the action

$$U|v_k\rangle = |v_{k+1}\rangle, \quad (4.45)$$

which implies that

$$U^m = \sum_{k=1}^{N} |v_{k+m}\rangle\langle v_k|. \quad (4.46)$$

The cyclic condition $|v_k + N\rangle = |v_k\rangle$ implies that $U^N = I$, and the eigenvalues $u_k = e^{i\,2\pi k/N} = \omega^k$ correspond to the eigenvectors $\langle u_k|$, with $|u_k + N\rangle = |u_k\rangle$ and $U|u_k\rangle = u_k|u_k\rangle$.

The operator V is defined by

$$V|u_k\rangle = |u_{k-1}\rangle, \quad (4.47)$$

implying that

$$V^n = \sum_{k=1}^{N} |u_{k-n}\rangle\langle u_k|. \tag{4.48}$$

It means that $V^n = I$ and that the operator V has eigenvalues $v_k = \omega^k$ and eigenvectors $|v_k\rangle$, namely, $V|v_k\rangle = v_k|v_k\rangle$. The relation

$$V^n U^m = \omega^{mn} U^m V^n \tag{4.49}$$

can be then derived from the cyclic condition $|v_k + N\rangle = |v_k\rangle$. Here $k, m, n, \ldots \in \mathbb{Z}_N$.

Eq. (4.49) is invariant under the substitutions $U \mapsto V$, $V \mapsto U^{-1}$, $m \mapsto n$, $n \mapsto -m$, leaving that Schwinger operators

$$S_{(m,n)} = \exp\left(i\frac{\pi}{N}mn\right) U^m V^n = \omega^{\frac{mn}{2}} U^m V^n, \tag{4.50}$$

invariant. These operators are also unitary. Besides,

$$S_{(0,0)} = I, \qquad S_{(m,n)}^{-1} = S_{(m,n)}^{\dagger} = S_{(-m,-n)}. \tag{4.51}$$

The set $S_{(m,n)}$ constitutes an orthonormal basis [Sch70] and also satisfy the properties:

1. associativity,

2. quasi-periodicity: $S_{(N,p)} = (-1)^p S_{(0,p)}$, $S_{(p,N)} = (-1)^p S_{(p,0)}$,

3. $S_{(N,N)} = (-1)^N S_{(0,0)}$,

4. $S_{(m,n)}^p = S_{(pm,pn)}$,

5. $\mathrm{tr}\big(S_{(m,n)}\big) = N\delta_{n0}\delta_{m0}$.

The matrix entries are governed by the rule

$$S_{(m,n)_{ij}} = \delta_{i-j,m}\omega^{n(i-m/2)}. \tag{4.52}$$

Besides, one can write

$$S_{(m,n)} = \sum_k e^{i(\pi/N)(2k+m)n}|v_{k+m}\rangle\langle v_k| \tag{4.53}$$

and an arbitrary matrix A can be written as

$$A = \frac{1}{N}\sum_{(m,n)} A_{(m,n)}S_{(m,n)}, \tag{4.54}$$

where $A_{(m,n)} = \mathrm{tr}\left(S^\dagger_{(m,n)}A\right)$.

Consider now a canonical basis E_{ij}, for $\{i,j\} \subset \{1,\ldots,n\}$, in the vector space of matrices, defined by

$$E_{ij}E_{kl} = \delta_{jk}E_{il}. \tag{4.55}$$

By defining the dual basis $\{E^{ij}\}$ in such a way that $E^{ij}(E_{kl}) = \delta^i_k\delta^j_l$, the coordinates are given by the trace $\mathrm{tr}_{ij} = \delta_{ij}$, namely, $\mathrm{tr} = \sum_{i=1}^N E^{ii}$.

The Schwinger and the canonical bases are related by

$$E_{ij} = \frac{1}{N}\sum_{n=1}^N \omega^{-n(i+j)/2}S_{(i-j,n)}, \tag{4.56}$$

$$S_{(m,n)} = \sum_{i=1}^N \omega^{n(i-m/2)} E_{i,i-m}. \tag{4.57}$$

When $N = 2$, the set of operators $S_{(m,n)}$ is led to the set of Pauli matrices.

A basis of great interest for the circulant matrices can be obtained when one takes $a_i = \delta_{1i}$ in (4.35). In this case the matrix

U is unitary, with entries $U_{ij} = \delta_{i,j+1}$. When U is composed with any another matrix, A, one obtains

$$(UA)_{ij} = A_{i-1,j}. \tag{4.58}$$

The left action by U dislocates the lines of A to one position lower, whereas the last line takes place of the first line of A. As A is a circulant matrix, this corresponds to $a_i \mapsto a_{i-1}$. Besides, the left action by U cyclically dislocates the columns of A by one position to the left. The set

$$\{U, U^2, U^3, \ldots, U^{N-1}, U^N = I\} \tag{4.59}$$

is a cyclic basis. The general expression (4.35) is simply a polynomial in U, and it is immediate to express

$$A = \sum_{i=1}^{N} a_i U^i. \tag{4.60}$$

In addition,

$$(U^m)_{ij} = \delta_{i,j+m}, \tag{4.61}$$

and the matrix U has eigenvalues ω^k and $(U^k_{diag})_{ij} = \delta_{ij}\omega^{ki}$. It is straightforward to verify that

$$U_{\text{diag}}U = \omega U U_{\text{diag}}, \tag{4.62}$$

where one denotes

$$U_{\text{diag}} = \text{diag}\left(\omega, \omega^2, \omega^3, \ldots, \omega^N\right). \tag{4.63}$$

Therefore (4.62) can be written as

$$VU = \omega UV, \tag{4.64}$$

implying that

$$V^m U^n = \omega^{nm} U^n V^m .$$ (4.65)

Such matrices U and V have crucial importance in the Weyl formulation of quantum mechanics [GP88].

In the basis $\{E_{ij}\}$, the cyclic basis for U and V are

$$U^j = \sum_{i=1}^{N} E_{i,N+i-j}, \qquad V^k = \sum_{i=1}^{N} \omega^{ik} E_{ii} .$$ (4.66)

The transformations

$$E_{ij} = \frac{1}{N} \sum_{n=1}^{N} \omega^{-nj} U^{i-j} V^n, \qquad U^i V^j = \sum_{s=1}^{N} \omega^{js} E_{s+i,s}$$ (4.67)

have Jacobian different of zero. Therefore any matrix $B = \sum_{i,j=1}^{N} B_{ij} E_{ij}$ can be written with respect to the basis $\{U^m V^n\}$ as

$$B = \frac{1}{N} \sum_{i,j=1}^{N} B_{ij} \omega^{-nj} U^{i-j} V^n .$$ (4.68)

Given an algebra with a set of generators $\{g_1, g_2, \ldots, g_{n-1}\}$ which satisfy

$$g_j^N = I,$$ (4.69)

$$g_k g_{k+1} = \omega \, g_{k+1} g_k,$$ (4.70)

$$g_k g_i = g_i g_k, \qquad \text{for } |k-1| \geq 2$$ (4.71)

then the set of elements $\{\sigma_i\}$ defined by

$$\sigma_i = \sum_{j=0}^{N-1} \omega^{j^2} g_i^j$$ (4.72)

do satisfy the braid relations. The problem now is to find generators satisfying conditions (4.69 – 4.71).

In the special case $N = 3$, the elements U and V do satisfy the requirement. Consider the operators

$$U = \begin{pmatrix} 0 & 0 & 1 \\ 1 & 0 & 0 \\ 0 & 1 & 0 \end{pmatrix}, \tag{4.73}$$

$$V = \begin{pmatrix} 1 & 0 & 0 \\ 0 & \omega & 0 \\ 0 & 0 & \omega^2 \end{pmatrix}, \tag{4.74}$$

$$\sigma_1 = I + \omega U + \omega^4 U^2 = \begin{pmatrix} 1 & \omega & \omega \\ \omega & 1 & \omega \\ \omega & \omega & 1 \end{pmatrix}, \tag{4.75}$$

$$\sigma_2 = I + \omega V + \omega^4 V^2 = \begin{pmatrix} 1 + 2\omega & 0 & 0 \\ 0 & 2 + \omega^2 & 0 \\ 0 & 0 & 2 + \omega^2 \end{pmatrix}. \tag{4.76}$$

Then one verifies directly that the braid relation

$$\sigma_1 \sigma_2 \sigma_1 = \sigma_2 \sigma_1 \sigma_2 \tag{4.77}$$

holds, consequently yielding a representation of the braid group B_3. An alternative point of view can be regarded. Define the operators $x = \sigma_1 \sigma_2 \sigma_1$ and $y = \sigma_2 \sigma_1$. It follows then, from the braid relation, that $x^2 = y^3$. This is a presentation of a knot group, the one of the trefoil knot.

Perhaps the simplest method to arrive at the Alexander polynomials makes use of a special relationship between knots and braids. A knot, if oriented and tame, can always be obtained as the closure of a braid. Its Alexander polynomial comes then from the Burau representation. It comes as a surprise that, at least in some cases, that polynomials also arise from Weyl operators.

Take $N = 2$ and the B_2 generator

$$\sigma_1 = I + \omega U = \begin{pmatrix} 1 & \omega \\ \omega & 1 \end{pmatrix}. \tag{4.78}$$

The trefoil comes from expressing

$$\sigma_1^3 = \begin{pmatrix} 1 + 3\omega^2 & 1 + 3\omega \\ 1 + 3\omega & 1 + 3\omega^2 \end{pmatrix} = \begin{pmatrix} 4 & 1 + 3\omega \\ 1 + 3\omega & 4 \end{pmatrix}. \tag{4.79}$$

If one puts $t = 3\omega$, it yields

$$\det(I - \sigma_1^3) = -(1 - t + t^2), \tag{4.80}$$

which is essentially the Alexander polynomial for the trefoil knot.

4.6 Potts model in terms of Weyl operators

Let us look for a realisation of the Potts model with M sites in terms of Weyl operators. The Weyl pairs of bounded operators include the (exponentiated) degrees of freedom and their conjugate momenta. Take M independent degrees of freedom (with Weyl operators U_j, V_k, for $\{i, k\} \subset \{1, 2, \ldots, M\}$), that is, one pair (U_j, V_k) on each site, but all operators corresponding to the same number N:

$$U_j^N = I, \qquad V_k^N = I, \qquad VU = e^{i\frac{2\pi}{N}} UV. \tag{4.81}$$

Notice that N regards the number of values spanned by the spin variable, which equals the number of assumable states at each site. Then one chooses the next-neighbor operators

$$
\begin{aligned}
u_1 &= V_1, & (4.82) \\
u_2 &= U_2^{-1}U_1, & (4.83) \\
u_3 &= V_2, & (4.84) \\
u_4 &= U_3^{-1}U_2, & (4.85) \\
u_5 &= V_3, & (4.86) \\
u_6 &= U_4^{-1}U_3. & (4.87)
\end{aligned}
$$

In general,

$$
u_{2i-1} = V_i, \qquad \text{and} \qquad u_{2i} = U_{i+1}^{-1}U_i. \tag{4.88}
$$

There are two interesting bases for the Hilbert space on which the Weyl operators act [Sch70]. In the so-called U-basis,

$$
\left\{ |k\rangle_U = |k_1, k_2, \ldots, k_M\rangle_U \right\}, \tag{4.89}
$$

the U_i are diagonal and the V_j are shift operators:

$$
\begin{aligned}
U_i|k\rangle_U &= e^{i\frac{2\pi}{N}k_i}|k_1, k_2, \ldots, k_i, \ldots, k_M\rangle_U, & (4.90) \\
V_i|k\rangle_U &= |k_1, k_2, \ldots, k_i - 1, \ldots, k_M\rangle_U. & (4.91)
\end{aligned}
$$

Notice that

$$
U_{i+1}^{-1}U_i|k_1, \ldots, k_M\rangle_U = e^{i\frac{2\pi}{N}(k_i - k_{i+1})}|k_1, \ldots, k_M\rangle_U, \tag{4.92}
$$

$$
\frac{1}{N}\sum_{i=1}^{N} U_{i+1}^{-1}U_i|k_1, \ldots, k_M\rangle_U = \delta_{k_i, k_{i+1}}|k_1, \ldots, k_M\rangle_U. \tag{4.93}
$$

One now defines the operators

$$p_k = \frac{1}{N} \sum_{j=1}^{N} u_k^j, \tag{4.94}$$

$$p_{2i-1} = \frac{1}{N} \sum_{j=1}^{N} V_i^j, \tag{4.95}$$

$$p_{2i} = \frac{1}{N} \sum_{j=1}^{N} U_{i+1}^{-j} U_i^j, \tag{4.96}$$

and find then that

$$p_{2i}|k\rangle_U = \delta_{k_i, k_{i+1}} |k\rangle_U, \tag{4.97}$$

$$p_{2i-1}|k\rangle_U = \frac{1}{N} \sum_{j=1}^{N} |k_1, \ldots, k_{i-1}, k_i - j, k_{i+1}, \ldots, k_M\rangle_U.$$

Notice that the $N = 2$ case yields

$$p_{2i-1} = \frac{1}{2} \left(I + V_i \right), \quad p_{2i} = \frac{1}{2} \left(I + U_{i+1} U_i \right) \tag{4.98}$$

and therefore

$$p_{2i-1}|k\rangle_U = \frac{1}{2}|k\rangle_U + \frac{1}{2}|k_1, \ldots, k_{i-1}, k_i - 1, k_{i+1}, \ldots, k_M\rangle_U,$$
$$p_{2i}|k\rangle_U = \delta_{k_i, k_{i+1}} |k\rangle_U. \tag{4.99}$$

The operators p_k generate a Jones algebra, A_{n+1}, which is a complex von Neumann algebra generated by the identity I and a set $\{p_1, \ldots, p_n\}$ of n projectors, satisfying the conditions

$$p_i^2 = p_i = p_i^\dagger, \tag{4.100}$$

$$p_i p_{i \pm 1} p_i = \tau p_i, \tag{4.101}$$

$$p_i p_j = p_j p_i, \quad \text{for } |i - j| \geq 2. \tag{4.102}$$

One finds that these operators have Jones index $\tau^{-1} = N$. The image is clear, where one has one N-valued degree of freedom per lattice site. Such algebras were known to physicists, as A_n, had essentially been used by Temperley and Lieb in their demonstration of the equivalence between the ice-type and the Potts models, the only difference being in the projector normalisation. The Temperley-Lieb projectors e_k are related to the projectors satisfying (4.100 – 4.102) by

$$e_k = \sqrt{N}p_k. \tag{4.103}$$

Then, with respect to the U-basis, one can express

$$e_{2i} = \frac{1}{\sqrt{N}} \sum_{j=1}^{N} U_{i+1}^{-j} U_i^j = \sqrt{N}\delta_{k_i,k_{i+1}} E^{\otimes M}, \tag{4.104}$$

$$e_{2i-1} = \frac{1}{\sqrt{N}} \sum_{j=1}^{N} V_i^j = E^{\otimes(i-1)} \otimes \left[\frac{1}{\sqrt{N}} \right] \otimes E^{\otimes(M-i)}, \tag{4.105}$$

where $\left[\frac{I}{\sqrt{N}} \right]$, which is in the ith position, is a diagonal $N \times N$ matrix with all the entries equal to $\frac{1}{\sqrt{N}}$. These are just the usual expressions used in the approach to the Potts model. Conditions (4.101) and (4.102) involve clearly a nearest neighbor prescription, and are reminiscent of the braid relations. Some linear combinations of the projectors and the identity do provide braid group generators:

$$G_i = \sqrt{t}\left[(1+t)p_i - I\right], \tag{4.106}$$

$$G_{2i-1} = \sqrt{t}\left(\frac{1+t}{N} \sum_{j=1}^{N} V_i^j - I \right), \tag{4.107}$$

$$G_{2i} = \sqrt{t}\left(\frac{1+t}{N} \sum_{j=1}^{N} U_{i+1}^{-j} U_i^j - I \right), \tag{4.108}$$

where the factors are only necessary to give the exact parametri-
sation for the skein relation.

The projectors are

$$p_i = \frac{G_i + \sqrt{t}}{(1+t)\sqrt{t}}, \tag{4.109}$$

and the condition $p_i^2 = p_i$ is equivalent to $G_i^2 = \sqrt{t}(t-1)G_i + t^2 I$,
or

$$\left(G_i - t\sqrt{t}I\right)\left(G_i + \sqrt{t}I\right) = 0, \tag{4.110}$$

or still

$$tG_i^{-1} - t^{-1}G_i + \frac{t-1}{\sqrt{t}}I = 0, \tag{4.111}$$

which is a skein relation.

⇨ **Comment 4.1.** ☞ For $N = 2$, the braid group generators are given by

$$G_{2i-1} = \sqrt{t}\left(\frac{t-1}{2}I + \frac{t+1}{2}V_i\right), \tag{4.112}$$

$$G_{2i} = \sqrt{t}\left(\frac{t-1}{2}I + \frac{t+1}{2}U_{i+1}U_i\right). \tag{4.113}$$

For the $N = 3$ case, they are given by

$$G_{2i-1} = \sqrt{t}\left[\frac{t-2}{3}I + \frac{t+1}{3}\left(V_i + V_i^2\right)\right], \tag{4.114}$$

$$G_{2i} = \sqrt{t}\left[\frac{t-2}{3}I + \frac{t+1}{3}\left(U_{i+1}^2 U_i + U_{i+1}U_i^2\right)\right], \tag{4.115}$$

and for $N = 4$ it follows that

$$G_{2i-1} = \sqrt{t}\left[\frac{t-3}{4}I + \frac{t+1}{4}\left(V_i + V_i^2 + V_i^3\right)\right], \tag{4.116}$$

$$G_{2i} = \sqrt{t}\left[\frac{t-3}{4}I + \frac{t+1}{4}\left(U_{i+1}^3 U_i + U_{i+1}^2 U_i^2 + U_{i+1}U_i^3\right)\right]. \tag{4.117}$$

One can express relations (4.107, 4.108) as

$$G_{2i-1} = \sqrt{t}\left[\frac{1+t}{N}\sum_{j=1}^{N-1}V_i^j + I\frac{t-(N-1)}{N}\right], \tag{4.118}$$

$$G_{2i} = \sqrt{t}\left[\frac{1+t}{N}\sum_{j=1}^{N-1}U_{i+1}^{N-j}U_i^j + I\frac{t-(N-1)}{N}\right]. \tag{4.119}$$

✓

One can extend the Jones algebra to J_∞, denoting a Jones algebra on infinitely many generators satisfying (4.100 – 4.102). J_n denotes the Jones algebra generated by an identity element I and the set of generators $\{p_1,\ldots,p_{n-1}\}$.

The relations (4.101 – 4.102) seem the braiding relations in the Artin braid group, which read

$$\begin{aligned}\sigma_i\sigma_{i\pm1}\sigma_i &= \sigma_{i\pm1}\sigma_i\sigma_{i\pm1}, & i &= 2,3,\ldots \\ \sigma_i\sigma_j &= \sigma_j\sigma_i, & |i-j| &> 1.\end{aligned} \tag{4.120}$$

This pattern led Jones to first construct a representation of the Artin braid group to his algebra, and then to discover an invariant of knots and links that is related to this representation.

The representation that Jones discovered is a linear one in the form

$$\begin{aligned}\rho : B_\infty &\to J_\infty \\ \sigma_i &\mapsto \rho(\sigma_i) = \alpha I + \beta e_i,\end{aligned}$$

for appropriate constants α and β.

To see how these representations work, it is useful to discuss the combinatorics of these algebras a bit further. The

Temperley-Lieb algebra, TL_n, is an algebra over a commutative ring $\mathbb{K}$ with generators $\{1, e_1, e_2, \ldots, e_{n-1}\}$ and relations [Kau871, Kau872, Kau88, Kau89]

$$e_i^2 = de_i, \tag{4.121}$$

$$e_i e_{i\pm 1} e_i = e_i, \tag{4.122}$$

$$e_i e_j = e_j e_i, \quad |i - j| > 1, \tag{4.123}$$

where $d \in \mathbb{K}$. These equations give the multiplicative structure of the algebra. The algebra is a free module over the ring $\mathbb{K}$, with basis given by the equivalence classes of these products modulo the given relations.

The concepts of Temperley-Lieb algebra and Jones algebra are interchangeable, in Ref. [Kau91]. Given a Jones algebra J_∞, with $p_i p_{i\pm 1} p_i = \tau p_i$, let $d = 1/\sqrt{\tau}$, assuming that this square root exists in the ring $\mathbb{K}$. Then one defines $e_i = dp_i$ and it follows that $p_i^2 = de_i$, with

$$
\begin{aligned}
e_i e_{i\pm 1} e_i &= \left(\frac{1}{\sqrt{\tau}}\right)^3 p_i p_{i\pm 1} p_i \\
&= \left(\frac{1}{\sqrt{\tau}}\right)^3 \tau p_i = \frac{p_i}{\sqrt{\tau}} \\
&= e_i, \tag{4.124}
\end{aligned}
$$

converting the Jones algebra to a Temperley-Lieb algebra.

4.7 Gases: symmetric group statistics

The case of a quantum ideal gas is very instructive, as it exhibits only those effects of purely statistical origin. The canonical partition function for a gas of N interacting particles has been given as an S_N-polynomial invariant. Interactions are represented by the cluster integrals b_k. The effect of statistics can be simulated by effective interactions. For a three-dimensional ideal quantum gas, it is enough to choose $b_k = \frac{(\pm 1)^{k-1}}{k^{5/2}}$, for bosons (upper sign) and fermions (lower sign). For a two-dimensional gas, the choice reads $b_k = \frac{(\pm 1)^{k-1}}{k^2}$.

Let us proceed to calculate the canonical partition function [Ald92]. We will refer to the standard 3-dimensional case only for comparison, and focus on a two-dimensional non-relativistic ideal gas. Momenta are then two-dimensional vectors, so that instead of the standard energy sphere, an energy circle. The number of microstates for particles on a surface of area S with energy less than or equal to E reads

$$\Sigma(E) = \frac{4\pi m S E}{h^2}, \tag{4.125}$$

and the corresponding number of microstates with energy between E and $E + dE$ is

$$g(E)dE = \frac{d\Sigma}{dE}dE = \frac{4\pi m S}{h^2}dE. \tag{4.126}$$

Unlike the three-dimensional case, the state density is constant here.

The canonical partition function for N particles is given by

$$Q_N(\beta, V) = \frac{S^N}{N!} \int d^2p_1 d^2p_2 d^2p_3 \ldots d^2p_N \times$$

$$e^{-\beta \sum_{i=1}^{N} \frac{p_i^2}{2m}} \langle p_1, p_2, \ldots, p_N | p_1, p_2, \ldots, p_N \rangle \tag{4.127}$$

(with $\beta = \frac{1}{kT}$). All the statistical content is contained in the normalisation amplitude $\langle p_1, p_2, \ldots, p_N | p_1, p_2, \ldots, p_N \rangle$. The state-representative ket $|p_1, p_2, \ldots, p_N\rangle$ is written as a sum of ordered products of one-particle kets $|p_j\rangle$ normalised as

$$\langle p_i | p_j \rangle = \delta^2(p_i - p_j), \tag{4.128}$$

and the resulting amplitude is a certain sum of deltas.

One can look at the first, second, third terms in each ordered product as respectively corresponding to the first, second, third particles, so that ultimately the physical ket is given as a sum of contributions of distinct particles, with coefficients fixed by statistics. For the $N = 2$ case, for example, the ket and its normalisation is, respectively, given by (upper sign for bosons, lower for fermions)

$$|p_1, p_2\rangle = \frac{1}{2}\Big[|p_1\rangle|p_2\rangle \pm |p_1\rangle|p_2\rangle\Big], \tag{4.129}$$

$$\langle p_1, p_2|p_1, p_2\rangle = \frac{1}{2}\Big[\delta^2(p_1-p_1)\delta^2(p_2-p_2) \pm \delta^2(p_1-p_2)\delta^2(p_2-p_1)\Big]. \tag{4.130}$$

This amplitude is a decomposition into cycles. The first factor gives the contributions of the two possible 1-cycles; the second, the contribution of the only 2-cycle. As the integrations in (4.127) put all momenta on an equal footing, it does not matter which momentum is in each place. Only the number of cycles of each type is important. One can introduce the notations

$$\hat{\delta}_1 = \delta^2(p_k - p_k), \tag{4.131}$$

for a 1-cycle contribution,

$$\hat{\delta}_2 = \delta^2(p_i - p_j)\delta^2(p_j - p_i), \tag{4.132}$$

for a 2-cycle contribution,

$$\hat{\delta}_3 = \delta^2(p_i - p_j)\delta^2(p_j - p_k)\delta^2(p_k - p_i), \tag{4.133}$$

for a 3-cycle contribution, and so on. In that case,

$$\langle p_1, p_2|p_1, p_2\rangle = \frac{1}{2}\Big[\hat{\delta}_1^2 \pm \hat{\delta}_2\Big]. \tag{4.134}$$

For $N = 3$, the ket

$$|p_1, p_2, p_3\rangle = \frac{1}{3!}\Big[|p_1\rangle|p_2\rangle|p_3\rangle \pm |p_2\rangle|p_1\rangle|p_3\rangle \pm |p_1\rangle|p_3\rangle|p_2\rangle$$
$$\pm |p_3\rangle|p_2\rangle|p_1\rangle + |p_2\rangle|p_3\rangle|p_1\rangle + |p_3\rangle|p_1\rangle|p_2\rangle\Big] \quad (4.135)$$

leads to

$$\langle p_1, p_2, p_3|p_1, p_2, p_3\rangle = \frac{1}{3!}\left[\hat{\delta}_1^3 \pm 3\hat{\delta}_1\hat{\delta}_2 + 2\hat{\delta}_3\right] . \quad (4.136)$$

The numerical factors just count the number of permutations of the corresponding cycle configuration. The amplitudes behave as cycle indicator polynomials:

$$\langle p_1, p_2, \ldots, p_N|p_1, p_2, \ldots, p_N\rangle = \frac{1}{N!}C_N\left[(\pm 1)^{j-1}\hat{\delta}_j\right] . \quad (4.137)$$

An amplitude is not a cycle indicator polynomial, but behaves like one, under the multiple integration sign in (4.127). That equation becomes, from (4.137),

$$Q_N(\beta, V) = \frac{1}{N!}C_N\left[\frac{(\pm 1)^{j-1}}{j}\frac{S}{\lambda^2}\right] . \quad (4.138)$$

This partition function is the sum of contributions of all configurations of distinct particles, with coefficients fixed by the statistics. In the present case, there is a one-to-one correspondence between such configurations and the elements of S_N. One can start from the configuration $|p_1\rangle|p_2\rangle|p_3\rangle \ldots |p_N\rangle$, corresponding to the identity element, and then get the remaining configurations by applying all the group elements. For this reason, the partition function for an ideal quantum gas will have a form analogous to that of a real classical gas, the statistical effects being simulated by an effective interaction represented by non-zero formal configuration integrals $b_k = \frac{(\pm 1)^{k-1}}{k^2}$.

The grand-canonical partition function, also known as the macro-canonical ensemble, is the statistical ensemble that is used to represent the possible states of a mechanical system of particles that are in thermal and chemical thermodynamic equilibrium with a reservoir. The grand-canonical partition function, written in terms of the fugacity $z = e^{\beta\mu}$, where here μ denotes the chemical potential, reads

$$\Xi(\beta, V, \mu) = \sum_{N=0}^{\infty} Q_N(\beta, V) z^N. \tag{4.139}$$

The expressions for the density,

$$n = \frac{N}{S} = z\frac{\partial}{\partial z}\log\Xi, \tag{4.140}$$

and for the pressure,

$$p = kT\log\Xi \tag{4.141}$$

involve, instead of the usual Bose and Fermi functions $g_{5/2}(z)$, $g_{3/2}(z)$, $f_{5/2}(z)$ and $f_{3/2}(z)$ of the three-dimensional case, the dilogarithm $g_2(z) = Li_2(z)$ and the logarithm, in the form $g_1(z) = -\log(1-z)$:

$$\frac{p\lambda^2}{kT} = \sum_{k=1}^{\infty} b_k z^k = \sum_{k=1}^{\infty} \frac{(\pm 1)^{k-1}}{k^2} z^k = \begin{cases} g_2(z) \\ f_2(z) = -g_2(-z), \end{cases} \tag{4.142}$$

$$\frac{N\lambda^2}{S} = \sum_{k=1}^{\infty} k b_k z^k = \sum_{k=1}^{\infty} \frac{(\pm 1)^{k-1}}{k} z^k = \begin{cases} g_1(z) \\ f_1(z) = -g_1(-z). \end{cases} \tag{4.143}$$

These relations are equivalent to

$$\frac{p\lambda^2}{kT} = \pm g_2(\pm z) = \pm Li_2(\pm z), \tag{4.144}$$

$$n\lambda^2 = \pm g_1(\pm z) = \mp \log(1 \mp z). \tag{4.145}$$

From this last equation, expressions for the fugacity can be obtained in terms of the degeneracy parameter $n\lambda^2$. For fermions,

$$z_f = e^{n\lambda^2} - 1, \tag{4.146}$$

whereas for bosons,

$$z_b = 1 - e^{-n\lambda^2}. \tag{4.147}$$

Notice that, for boson and fermion gases of same temperature, same mass and same value of $n\lambda^2$, the expressions

$$z_f = \frac{z_b}{1 - z_b} \quad \text{and} \quad z_b = \frac{z_f}{1 + z_f} \tag{4.148}$$

hold.

The equations of state are, for bosons and fermions, respectively

$$\left[\frac{p}{kT}\right]_b = \frac{1}{n\lambda^2} g_2\left(1 - e^{-n\lambda^2}\right); \tag{4.149}$$

$$\left[\frac{p}{kT}\right]_f = -\frac{1}{n\lambda^2} g_2\left(1 - e^{n\lambda^2}\right). \tag{4.150}$$

Bosons exhibit, in the three-dimensional case, the very special phenomenon of Bose-Einstein condensation. Below a certain critical temperature, a macroscopic number of particles stand in the fundamental, least energy state, to constitute the condensate. The best indication of this critical phenomenon is given by the behavior of the specific heat at constant volume, whose derivative becomes infinite at the critical temperature. This was first seen in liquid He^4. In a liquid, of course, the molecules interact with each other actively. It is, nevertheless, a nice signal for condensation that a three-dimensional boson gas shows already a similar behavior even in the absence of interactions.

Condensation appears then as an effect of statistics. The specific heat at constant volume has, near the critical temperature, a form resembling a lambda (Λ) shape. Hence the name "λ-point" is frequently given to the critical point. The relation to superconductivity[2] comes from the tendency of electrons at very low temperatures to correlate into bosonic pairs, which can then condensate. Notice, finally, that the specific heat measures energy fluctuations

$$kT^2 C_V = \langle E^2 \rangle - \langle E \rangle^2. \tag{4.151}$$

Its singular behavior indicates, consequently, the presence of high energy oscillations.

One should examine the specific heat (here at constant surface instead of volume), to verify whether also in two dimensions an ideal quantum gas exhibits a λ-point. Notice, to begin with, that the internal energy is given by $U = PS$. The constant-surface specific heat reads

$$C_S = \left(\frac{\partial U}{\partial T} \right)_{N,S}. \tag{4.152}$$

A direct calculation shows that

$$\left(\frac{C_S}{Nk} \right)_{\text{bosons}} = \frac{2}{n\lambda^2} g_2 \left(1 - e^{-n\lambda^2} \right) - \frac{n\lambda^2}{e^{n\lambda^2} - 1}. \tag{4.153}$$

---

[2]Superconductivity is a state of some materials where electrical resistance equals zero, whereas the magnetic flux fields are let out of the material. Unlike metallic conductors, whose resistance decreases gradually as its temperature is lowered even down to near absolute zero Kelvin, a superconductor has a characteristic critical temperature below which the resistance drops abruptly to zero. An electric current through a loop of superconducting wire can persist indefinitely with no power source [BCS57].

Quantum effects are to be expected for large values of $n\lambda^2$. One verifies that no derivative singularity occurs in this two-dimensional case. The ground state occupancy reads

$$(N_0)_{\text{bosons}} = \frac{z_b}{1 - z_b} = e^{n\lambda^2} - 1. \qquad (4.154)$$

For lower and lower temperatures the ground state gets more and more crowded, but with no singularity.

There is, in fact, a curious point: if one calculates the fermionic case, use (4.148) and the property of the dilogarithm $g_2(1 - x) + g_2(1 - \frac{1}{x}) = -\frac{1}{2}(\log x)^2$, it yields

$$\left(\frac{C_S}{Nk}\right)_{\text{fermions}} = \left(\frac{C_S}{Nk}\right)_{\text{bosons}}. \qquad (4.155)$$

Bosons and fermions have, consequently, the same response to local energy fluctuations. It is possible to show that the same is not true for fluctuations in the number of particles. The main point is that ideal gases, in two dimensions, exhibit no λ-point.

We will see in Sect. 4.9 that a braid gas interpolates between a boson and a fermion gas, and that occurs in a very interesting way. To obtain the basic B_N kets, however, it will be convenient to discuss beforehand the meaning of the S_N kets used above.

4.8 Covering spaces

To see what happens when exchange groups distinct of S_N are at work, one needs a more detailed understanding of the meaning of the decompositions like (4.129) and (4.135). The configuration space for N identical interpenetrable particles is E^{2N}/S_N. The space E^{2N} is the universal covering, as the fundamental group $\pi_1\left(E^{2N}\right)$ is just S_N.

A covering of a space X is another space that is locally home-omorphic to X, consisting of an unfolding of X which breaks some equivalence between its points. Every space has a unique universal covering, which is simply-connected, meaning that the first fundamental group, π_1, equals the identity and whose folds, or sheets, have a one-to-one relationship with the elements of π_1. To go on, one must present monodromy. This important concept, related to covering maps and their degeneration into ram-ification, starts on the study of certain functions that fail to be single-valued as one runs around a path encircling a singularity. The failure of monodromy can be measured when one defines the monodromy group. The different values of a multivalued function Ψ on a multiply-connected space are obtained through a representation of a group, the monodromy group of Ψ, in gen-eral, a subgroup of π_1. A function becomes single-valued on a covering whose sheets are in a one-to-one relationship with the elements of its monodromy group. All functions become single-valued on the universal covering.

⇨ **Comment 4.2.**   Complex analysis is a fertile field of examples regarding monodromy. A function that is an analytic function $F(z)$ in some open subset E of the punctured complex plane $\mathbb{C} \setminus \{0\}$ can be continued back into E, with different values. For example, if one takes

$$
\begin{aligned}
F(z) &= \log(z), \\
E &= \{z \in \mathbb{C} \mid \mathrm{Re}(z) > 0\},
\end{aligned}
\tag{4.156}
$$

then analytic continuation anti-clockwise round the circle $|z| = 1$ will not yield $F(z)$ but $F(z) + 2\pi i$, instead. In this case the monodromy group is infinite cyclic and the covering space is the universal cover of the punctured complex plane. This cover can be visualised as the helicoid

$$
\begin{aligned}
x &= \rho \cos(a\theta), \\
y &= \rho \sin(a\theta), \\
z &= \theta,
\end{aligned}
\tag{4.157}
$$

where α is a constant that defines the helicoid chirality and $\rho > 0$. The covering mapping is a vertical projection, in a sense collapsing the spiral in the natural way to get a punctured plane. ✓

Consider, to fix the ideas, the $N = 2$ case. Suppose that positions are enough to describe the particles, and call x_1 and x_2 the position vectors of the first and the second particles, respectively. The covering space $\mathbb{E}^4$ is the set $\{(x_1, x_2)\}$. The physical configuration space X would be the same, but with points (x_1, x_2) and (x_2, x_1) identified. The point (x_2, x_1) is obtained from (x_1, x_2) by the action of the transposition

$$
\begin{aligned}
s_1 : X &\rightarrow X \\
(x_1, x_2) &\mapsto s_1(x_1, x_2) = (x_2, x_1).
\end{aligned}
\tag{4.158}
$$

A complex function $\Psi(x_1, x_2)$, describing the wave function of the two-particle system, is single-valued on the covering space, but two-valued on the configuration space. In fact, the space $\mathbb{E}^4/S_2$ involves a cone [LM77] and is rather difficult to picture out. To make a drawing easier to look at, one considers instead the covering related to the function $\sqrt{z}$, whose group, the cyclic group $\mathbb{Z}_2$, is isomorphic to S_2. The scheme in Fig. 47 shows how $\Psi(x_1, x_2)$ is single-valued on $\mathbb{E}^4$, where $(x_2, x_1) \neq (x_1, x_2)$, and double-valued on $\mathbb{E}^4/S_2$, where the two values $\Psi(x_1, x_2)$ and $\Psi(x_2, x_1)$ correspond to the same point $(x_2, x_1) \sim (x_1, x_2)$.

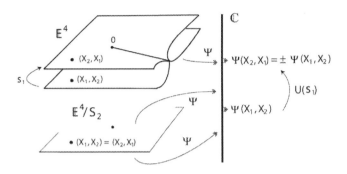

Figure 47: Qualitative representation of the two-particle configuration space: S_2 statistics.

Now, the expression

$$\begin{aligned}
\Psi(x_2, x_1) &= \Psi(s_1(x_1, x_2)) = U(s_1)\Psi(x_1, x_2) \\
&= \pm \Psi(x_1, x_2),
\end{aligned} \tag{4.159}$$

can obtained from $\Psi(x_1, x_2)$ by the action of an operator $U(s_1)$ representing s_1 on the Hilbert space of wave functions. There are two sheets, because s_1 applied twice is the identity. Commonly used wave functions are taken on the covering space, where they are single-valued, and on which particles are supposed to be distinguishable. Here the monodromy group is the whole group S_2 and two sheets appear, one for each distinct group element, 1 and s_1. This means that in (4.129) one sums over all distinct sheet contributions.

An analogous treatment can be applied to the $N = 3$ case (4.135), in which $6 = 3!$ contributions come out, since it is the number of elements of S_3, one for each sheet. The academic example of the $\mathbb{Z}_N$-gas mentioned, would show no difference in the $N = 2$ case, since the groups $\mathbb{Z}_2$ and S_2 are isomorphic, but would exhibit quite a different covering for $N = 3$, as the monodromy group $\mathbb{Z}_3$ would require three sheets. One learns in this way what is really accomplished when physical kets or wave functions are written in terms of distinct-particle contributions. Indeed, a superposition of all the values is taken. We will see in Sect. 4.9 that braid statistics does require infinitely-folded covering spaces, yet normalisation eliminates all but two of the infinite contributions.

4.9 Gases: braid group statistics

Manifestations of exotic statistics are to be expected as long as impenetrable particles are considered in two dimensions [Wu92], whatever their internal structure is. Braid statistics has been an object of highly sophisticated analysis [FG90]. We intend here only to examine an ideal gas obeying such statistics. Such a free gas might come to model, up to interactions, the electron gas in superconductors, in which the electrons are confined to the surface. It is anyhow the starting point, to which dynamical effects are to be added in a deeper analysis. Dynamics has been mainly studied in the line of Landau-Ginzburg models.

As comprised in the discussion subsequent to Eq. (4.127), all the statistical content lies in the normalisation amplitude $\langle p_1, p_2, \ldots, p_N | p_1, p_2, \ldots, p_N \rangle$. In order to obtain the convenient representations for the physical kets in terms of ordered products of one-particle kets, analogous to (4.129) and (4.135), one must proceed to an analysis similar to that of Fig. 47.

Take again the $N = 2$ case. The covering space has now infinite sheets, as illustrated in Fig. 48. The physical ket has the general form

$$|p_1, p_2\rangle = f(\phi) \, |p_1\rangle |p_2\rangle + g(\phi) \, |p_2\rangle |p_1\rangle. \qquad (4.160)$$

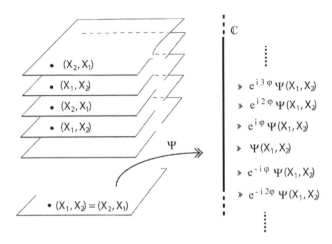

Figure 48: The infinite unfolding of the two-particle configuration space for braid statistics.

Contributions along $|p_1\rangle|p_2\rangle$ come from all colored elements, those which ultimately do not exchange the particles, such as 1, σ_1^2, σ_1^{-2}, σ_1^4, σ_1^{-4} ..., σ_1^{2n}, σ_1^{-2n}. For instance, one can take

$$f(\phi) = 1 + e^{i2\phi} + e^{-i2\phi} + e^{i4\phi} + e^{-i4\phi} + \cdots = \sum_{k\in\mathbb{Z}} e^{2ik\phi}. \quad (4.161)$$

As to $g(\phi)$, it receives all particle exchanging contributions, those coming from the odd powers of σ_1, given by

$$g(\phi) = e^{i\phi} + e^{-i\phi} + e^{i3\phi} + e^{-i3\phi} + \cdots = e^{i\phi} f(\phi). \quad (4.162)$$

There is an arbitrariness in the choice of the initial sheet, that one corresponding to the identity element in the infinite foliation. This arbitrariness is reflected in the indeterminacy of the series for both $f(\phi)$ and $g(\phi)$. It is always possible to choose for

$f(\phi)$ some real though indeterminate series keeping with $g(\phi)$ the relation

$$g(\phi) = e^{i\phi} f(\phi). \qquad (4.163)$$

In consequence, $|p_1, p_2\rangle = f(\phi)\,|p_1\rangle|p_2\rangle + e^{i\phi} f(\phi)\,|p_2\rangle|p_1\rangle$, which can be normalised to become

$$|p_1, p_2\rangle = \frac{1}{2}\left[|p_1\rangle|p_2\rangle + e^{i\phi}|p_2\rangle|p_1\rangle\right] \qquad (4.164)$$

Then, the indeterminacy has been eliminated. Of course, this reduces to (4.129), when $\phi = 0$ and π, and one falls back into the penetrability case $\sigma_1^2 = 1$. For $N = 3$, an analogous, though far more involved, analysis leads to

$$|p_1, p_2, p_3\rangle = \frac{1}{3!}\Big[|p_1\rangle|p_2\rangle|p_3\rangle + e^{i\phi}|p_2\rangle|p_1\rangle|p_3\rangle + e^{i\phi}|p_1\rangle|p_3\rangle|p_2\rangle$$
$$+ e^{i\phi}|p_3\rangle|p_2\rangle|p_1\rangle + e^{i2\phi}|p_2\rangle|p_3\rangle|p_1\rangle + e^{i2\phi}|p_3\rangle|p_1\rangle|p_2\rangle\Big], \quad (4.165)$$

generalizing (4.135). In this way a realisation of the physical kets, in terms of products of distinguished-particle kets, is thus obtained, with the symmetric group still selecting the terms. The coefficients are, after normalisation, simply products of terms $e^{i\phi}$ corresponding to the number of transpositions. Because such a realisation is still feasible, the symmetric group keeps a fundamental role and its general consequences, as the cycle decompositions and indicator polynomials, remain valid.

Eqs. (4.164 – 4.165) yield

$$\langle p_1, p_2 | p_1, p_2 \rangle = \frac{1}{2}\left(\hat{\delta}_1^2 + \cos\phi\,\hat{\delta}_2\right), \qquad (4.166)$$

$$\langle p_1, p_2, p_3 | p_1, p_2, p_3 \rangle = \frac{1}{3!}\left(\hat{\delta}_1^3 + 3\cos\phi\hat{\delta}_1\hat{\delta}_2 + 2\cos^2\phi\hat{\delta}_3\right) (4.167)$$

The sum over distinct classical configurations is restored, with a generalisation of the effective statistical interaction: concerning the boson and fermion cases, the signals are replaced by $\cos\phi$. Therefore the purely combinatorial aspects remain the same as for the S_N group, the amplitudes keeping their cycle decomposition character. Examination of higher values of N leads to a simple rule,

> ☞ *to obtain the formal configuration integrals, starting from those ones for the symmetric group, it suffices to make the substitution* $(\pm 1)^{j-1} \mapsto \cos^{j-1}\phi$.

The canonical partition function is, instead of (4.138), a straightforward generalisation of it, given by

$$Q_N(\beta, V) = \frac{C_N}{N!}\left(\frac{\cos^{j-1}\phi}{j}\frac{S}{\lambda^2}\right). \qquad (4.168)$$

From that, one can proceed to thermodynamics, where the relations [Ald92]

$$\frac{p}{kT} = \frac{S}{\lambda^2}\sec\phi\, g_2(z\cos\phi), \qquad (4.169)$$

$$n = \frac{1}{\lambda^2}\sec\phi\, g_1(z\cos\phi), \qquad (4.170)$$

$$n\lambda^2\cos\phi = g_1(z\cos\phi) = -\log(1 - z\cos\phi), \quad (4.171)$$

hold, then implying that

$$z\cos\phi = 1 - e^{-n\lambda^2\cos\phi}.$$

The ground state occupancy turns up to be

$$N_0 = \frac{z}{1 - z\cos\phi}. \qquad (4.172)$$

Condensation only appears in the bosonic case, but a high ground-state concentration of particles can be attained whenever $\cos\phi$ approaches the value $+1$. The equation of state appears as

$$\frac{p}{nkT} = \frac{1}{n\lambda^2\cos\phi}\, g_2\left(1 - e^{-n\lambda^2\cos\phi}\right). \qquad (4.173)$$

This includes as extreme cases both the bosonic ($\cos\phi = 1$) and the fermionic ($\cos\phi = -1$) cases. Besides, also the Boltzmann case is included, when $\cos\phi = 0$. The equation of state changes progressively with the value of $\cos\phi$, with the remarkable intermediate classical case $\frac{p}{nkT} = 1$. All the physical quantities interpolate in a very simple way those of S_N statistics. The internal energy $U = -\left[\frac{\partial}{\partial\beta}\log\Xi\right]_{z,S}$ and the specific heat $C_S = \left[\frac{\partial U}{\partial T}\right]_{N,S}$ are of special interest, as one is interested in the eventual presence of a lambda point:

$$U = pS = \frac{S\,kT}{\lambda^2\cos\phi}\,g_2\left(1 - e^{-n\lambda^2\cos\phi}\right), \qquad (4.174)$$

$$\left[\frac{C_S}{Nk}\right]_\phi = \frac{2}{n\lambda^2\cos\phi}\,g_2\left(1 - e^{-n\lambda^2\cos\phi}\right) - \frac{n\lambda^2\cos\phi}{e^{n\lambda^2\cos\phi} - 1}. \qquad (4.175)$$

The symmetries found in the S_N case reappear here. In fact, the term $\left[\frac{C_S}{Nk}\right]_\phi$ is the same for $\cos\phi$, with opposite signs. As $|\cos\phi|$ tends to zero, the two-dimensional Dulong-Petit limit,

$$\left[\frac{C_S}{Nk}\right]_{\pi/2} = 1, \qquad (4.176)$$

is approached.

No sign of λ-point shows up, even in the bosonic case. Consequently, even if some condensation comes to take place, no abrupt transition is to be expected. Starting from the fermionic case, the specific heat is continuously deformed as $\cos\phi$ tends to zero, reaches a straight horizontal line at this limit, and then retraces back its way down to the bosonic case, identical to the starting point.

There is a different statistics for each value of the angular parameter ϕ, which is, in principle, totally arbitrary. Notice, however, that $\phi = 0$ and $\phi = \pi$ correspond to penetrable particles. On the other hand, braid gases interpolate between bosons and fermions in such a way that Boltzmann particles stay in the middle, $\phi = \pi/2$, a curious intermediate case with distinguishable, classical particles. Taking some risk in forwarding an interpretation, one might take ϕ as a measure of penetrability and, consequently, of topological puncturedness. Indeed, the more ϕ departs from the extreme values, the less are the particles allowed to penetrate each other, utmost impenetrability standing in the middle. Quantum effects (like degeneracy) come precisely from forced superposition of individual particle wave functions. With highest impenetrability forbidding such superposition, it would be natural to find it related to classical behavior.

As a final point, let us recall that, in the case of usual superconductivity, it was London's remark about the lambda structure in ideal boson gas which triggered the idea that some kind of bosonisation played a fundamental role in the phenomenon. Notice that the absence of a λ-point by no means excludes the possibility of phase transitions in real cases, when dynamics become dominant. In reality, there seems to be a good theoretical in favour of its presence.

Exercises

(**1**) Consider the one-dimensional Ising model

$$H = -J \sum_{i=1}^{N} s_i s_{i+1}, \tag{4.177}$$

with the periodic boundary condition $s_N + 1 = s_1$. Show that the canonical partition function Z at inverse temperature β can be expressed as [Hil13]

$$Z = e^{NG} \operatorname{tr}\left(T^N\right), \tag{4.178}$$

where $T = e^{\mathcal{J}^*} \sigma^x$ is the transfer matrix, being σ^x the first Pauli spin matrix. Find G and $\mathcal{J}^*$ in terms of $\mathcal{J}^* = \beta J$ and determine Z from (4.185).

(2) Consider the Ising model on the ladder lattice,

$$H = -J \sum_{i=1}^{N} \left(s_{i,1} s_{i,2} + s_{i,1} s_{i+1,1} + s_{i,2} s_{i+1,2} \right), \qquad (4.179)$$

with the periodic boundary condition $s_{N+1,j} = s_{1,j}$, for $j = 1, 2$. Show that the canonical partition function Z at inverse temperature β can be expressed as [Hil13]

$$Z = e^{2NG} \operatorname{tr} \left(T^N \right), \qquad (4.180)$$

where the 4×4 transfer matrix, T, reads

$$T = UV, \qquad (4.181)$$

where

$$U = e^{\mathcal{J}} \sigma_1^z \sigma_2^z \qquad (4.182)$$
$$V = e^{\mathcal{J}^*} \left(\sigma_1^x + \sigma_2^x \right), \qquad (4.183)$$

where σ^z is the third Pauli spin matrix.

(3) Consider the Ising model on a $N \times M$ lattice,

$$H = -J \sum_{i=1}^{N} \sum_{j=1}^{M} \left(s_{i,j} s_{i+1,j} + s_{i,j} s_{i,j+1} \right), \qquad (4.184)$$

with the periodic boundary condition $s_{N+1,j} = s_{1,j}$ and $s_{i,M+1} = s_{i,1}$. Show that [Hil13]

$$Z = e^{NMG} \operatorname{tr} \left(T^N \right), \qquad (4.185)$$

where the $2^M \times 2^N$ transfer matrix, T, reads

$$T = UV, \tag{4.186}$$

where

$$
\begin{aligned}
U &= e^{\mathcal{U}} & (4.187) \\
V &= e^{\mathcal{V}}. & (4.188)
\end{aligned}
$$

Compute $\mathcal{U}$ and $\mathcal{V}$ with respect to the sets of Pauli spin matrices $\{\sigma_1^z, \ldots, \sigma_M^z\}$ and $\{\sigma_1^x, \ldots, \sigma_M^x\}$, respectively.

Bibliography

[AB26] J. Alexander, G. Briggs, *On types of knotted curves*, Ann. Math. **28** (1926) 562.

[Ada01] C. C. Adams, "The Knot Book", W. H. Freeman, New York, 2001.

[Ada78] J. F. Adams, *Infinite loop spaces*, Annals of Mathematics Studies **90**, Princeton Univ. Press, Princeton, 1978.

[AGR14] R. Abłamowicz, I. Gonçalves and R. da Rocha, *Bilinear Covariants and Spinor Fields Duality in Quantum Clifford Algebras*, J. Math. Phys. **55** (2014) 103501 [arXiv:1409.4550 [math-ph]].

[AW88] Y. Akutsu, M. Wadati, *Knots, Links, Braids and Exactly Solvable Models in Statistical Mechanics*, Commun. Math. Phys. **117** (1988) 243.

[Ald92] R. Aldrovandi, *Two-dimensional quantum gas*, Fortschr. Physik **40** (1992) 631.

[Ale28] J. W. Alexander *Topological invariants of knots and links*, Transactions of the American Math. Soc. **30** (1928) 275.

[AP95] R. Aldrovandi and J. G. Pereira, "An Introduction to Geometrical Physics", World Scientific, Singapore, 1995.

[Ald01] R. Aldrovandi, "Special Matrices of Mathematical Physics", World Scientific, Singapore, 2001.

[And66] G. E. Andrews, *q-Series: Their development and application in analysis, number theory, combinatorics, physics and computer algebra*, C.B.M.S. Regional Conference Series in Math. **66** AMS, Providence, 1986.

[AAR99] G. E. Andrews, R. Askey, R. Roy, "Special Functions", Cambridge Univ. Press, Cambridge, 1999.

[Arm02] M. Armstrong, "Basic Topology", Undergraduate Texts in Mathematics. Springer-Verlag, New York, Berlin, Heidelberg, 1983.

[Art47] E. Artin, *Theory of braids*, Annals of Math. **48** (1947) 101.

[Asc06] P. Aschieri, "Lectures on Hopf Algebras, Quantum Groups and Twists", Lectures given at the second Modave Summer School in Mathematical Physics, August 6-12, 2006 [arXiv:hep-th/0703013].

[Ash92] A. Ashtekar, C. Rovelli and L. Smolin, *Weaving a classical geometry with quantum threads*, Phys. Rev. Lett. **69** (1992) 237.

[Ash93] C. W. Ashley, "The Ashley Book of Knots", International Guild of Knot Tyers by Doubleday, New York, 1993.

[Ati91] M. F. Atiyah, "The Geometry and Physics of Knots", Cambridge University Press, Cambridge, 1991.

[Bae94] J. Baez and J. P. Muniain, "Gauge Fields, Knots and Gravity", Series on Knots and Everything: Volume 4, World Scientific, London, 1994.

[Ban09] P. D. Bangert, *Braids and Knots*, in R. Ricca (ed.) "Lectures on Topological Fluid Mechanics", Lecture Notes in Mathematics **1973**, Springer, Berlin, 2009.

[Bax82] R. J. Baxter, "Exactly Solved Models in Statistical Mechanics", Academic Press, London, 1982.

[BB11] L. Bonora and A. A. Bytsenko, *Partition fnctions for quantum gravity, black holes, elliptic genera and lie algebra homologies,* Nucl. Phys. B **852** (2011) 508 [arXiv:1105.4571 [hep-th]].

[BBG16] L. Bonora, A. A. Bytsenko and A. E. Goncalves, *Chern-Simons invariants on hyperbolic manifolds and topological quantum field theories,* Eur. Phys. J. C **76** (2016) no.11, 625 [arXiv:1606.02554 [hep-th]].

[BCS57] J. Bardeen, L. Cooper, J. R. Schriffer, *Theory of superconductivity,* Phys. Review **108** (1957) 1175.

[Ben99] H. B. Benaoum, *(q, h)-analogue of Newton's binomial formula,* J. Phys. A: Math. Gen. **32** (1999) 2037.

[Bir74] J. S. Birman, "Braids, Links and Mapping Class Groups", Ann. of Math. Studies **82**, Princeton Univ. Press, Princeton, 1974.

[Bir91] J. S. Birman, *Recent developments in braid and link theory*, Math. Intell. **13** (1991) 52.

[Bir93] J. S. Birman, *New points of view in knot theory*, Bull. Am. Math. Soc. **28** (1993) 253.

[Bir75] J. S. Birman, "Braids, Links, and Mapping Class Groups", Princeton Univ. Press, Princeton, 1975.

[BKL98] J. S. Birman, K. H. Ko, S. J. Lee, *A new approach to the word and conjugacy problems in the braid groups*, Adv. Math. **139** (1998) 322.

[BN95] D. Bar-Natan, *On the Vassiliev knot invariants*, Topology **34** (1995) 423.

[Boh47] F. Bohnenblust, *The algebraical braid group*, Ann. Math. **48** (1947) 127.

[BO18] G. P. de Brito, N. Ohta, A. D. Pereira, A. A. Tomaz and M. Yamada, *Asymptotic safety and field parametrization dependence in the $f(R)$ truncation*, Phys. Rev. D **98** (2018) 026027 [arXiv:1805.09656 [hep-th]].

[BR07] A. E. Bernardini and R. da Rocha, *Lorentz-violating dilatations in the momentum space and some extensions on non-linear actions of Lorentz algebra-preserving systems*, Phys. Rev. D **75** (2007) 065014 [arXiv:hep-th/0701094 [hep-th]].

[BR08] A. E. Bernardini and R. da Rocha, *Obtaining the equation of motion for a fermionic particle in a generalized Lorentz-violating system framework*, EPL **81** (2008) 40010 [arXiv:hep-th/0701092 [hep-th]].

[BR12] A. E. Bernardini and R. da Rocha, *Dynamical dispersion relation for ELKO dark spinor fields*, Phys. Lett. B **717** (2012) 238 [arXiv:1203.1049 [hep-th]].

[Bur36] W. Burau, *über Zopfgruppen und gleichsinnig verdrillte Verkettungen*, Abh. Math. Sem. Univ. Hamburg **11** (1936) 179.

[BX93] J. Birman and X.-S. Lin, *Knot polynomials and Vassiliev's invariants*, Inv. Mathematicae **111** (1993) 225.

[CF63] R. H. Crowell, R. H. Fox, "Introduction to Knot Theory", Springer, Berlin, 1963.

[CMP98] S. Cho, J. Madore, K. S. Park, *Non-commutative geometry of the h-deformed quantum plane*, J. Phys. A: Math. Gen. **31** (1998) 2639.

[Com74] L. Comtet, "Advanced Combinatorics", Reidel, Dordrecht, 1974.

[Con70] J. H. Conway, *An Enumeration of Knots and Links, and Some of Their Algebraic Properties*, In J. Leech (editor), "Computational Problems in Abstract Algebra", Pergamon Press, Oxford, pp. 329-358, 1970.

[Con90] J. H. Conway, J. C. Lagarias, *Tilings with polyminoes and combinatorial group theory*, J. Comb. Theor. A **53** (1990) 183.

[Con94] A. Connes, "Noncommutative Geometry", Academic Press, Boston, 1994.

[CT92] F. Constantinescu and F. Toppan, *On the linearized Artin braid representation* [arXiv:hep-th/9210020 [hep-th]].

[DNF79] B. Doubrovine, S. Novikov & A. Fomenko, "Géométrie Contemporaine", MIR, Moscow, 1979.

[DS06] D. S. Silver, *Knot theory's odd origins*, American Scientist **94** (2006) 158.

[Fad96] L. D. Faddeev and A. J. Niemi, *Knots and particles*, Nature **387** (1997) 58 [hep-th/9610193].

[FG90] J. Frohlich and F. Gabbiani, *Braid statistics in local quantum theory*, Rev. Math. Phys. **2** (1991) 251.

[FGP96] H. Fort, R. Gambini and J. Pullin, *Lattice knot theory and quantum gravity in the loop representation*, Phys. Rev. D **56** (1997) 2127 [arXiv:gr-qc/9608033 [gr-qc]].

[Fis00] L. Fischer, *Drawing knot pictures using LATEX with XY-pic — An introduction* [http://www.lars.fischer.de.vu].

[FK89] J. Fröhlich, C. King, *Two-dimensional conformal field theory and three-dimensional topology*, Int. J. Mod. Phys. A **22** (1989) 5321.

[FKW02] M. H. Freedman, A. Kitaev, Z. Wang, *Simulation of topological field theories by quantum computers*, Comm. Math. Phys. **227** (2002) 587.

[Fra74] J. B. Fraleigh, "A First Course in Abstract Algebra", Addison–Wesley, Reading, Mass., 1974.

[FRS16] A. F. Ferrari, R. da Rocha and J. A. Silva-Neto, *The role of singular spinor fields in a torsional gravity, Lorentz-violating, framework,* Gen. Rel. Grav. **49** (2017) 70 [arXiv:1607.08670 [hep-th]].

[FYH85] P. Freyd, D. Yetter, J. Hoste, W. B. R. Lickorish, K. Millet, A. Ocneanu, *A new polynomial invariant of knots and links*, Bull. Amer. Math. Soc. **12** (1985) 239.

[GP74] V. Guillemin, A. Pollack, "Differential Topology", Prentice Hall, Englewood Cliffs, 1974.

[GP88] D. Galetti and A. F. R. Toledo Piza, *An extended Weyl-Wigner transformation for special finite spaces*, Physica A **149** (1988) 267.

[Hak02] D. Hakon et al, *Tópicos em Matemática Quântica*, 22th Colóquio Brasileiro de Matemática, IMPA, Rio de Janeiro 2002.

[Ham62] M. Hamermesh, "Group Theory and its Applications to Physical Problems", Addison–Wesley, Reading, Mass., 1962.

[Hat02] A. Hatcher, "Algebraic Topology", Cambridge Univ. Press, Cambridge, 2002.

[HBC20] J. M. Hoff da Silva, D. Beghetto, R. T. Cavalcanti and R. da Rocha, *Exotic fermionic fields and minimal length*, Eur. Phys. J. C **80** (2020) 727 [arXiv:2006.03490 [hep-th]].

[Hel58] H. Helmholtz, *Über Integrale der hydrodynamischen Gleichungen, welche den Wirbelbewegungen entsprechen*, J. Reine Angew. Math. **55** (1858) 25.

[Hel67] H. Helmholtz, *On integrals of the hydrodynamical equations, which express vortex-motion* (Translated by P. G. Tait), Phil. Mag., Ser. 4, **33** (1867) 485.

[Hel69] H. Helmholtz, *Sui movimenti dei liquidi*, Nuovo Cimento, Ser. 2, **1** (1869) 289.

[Hil13] H. Hilhorst, *Advanced topics in statistical physics*, http : //www.th.u − psud.fr/page_perso/Hilhorst/N/ex13I.pdf, Laboratoire de Physique Théorique, Université de Paris-Sud, France, 2013.

[HTW98] J. Hoste, M. Thistlethwaite, J. Weeks, *The first 1,701,936 knots*, Math. Intelligencer **20** (1998) 33.

[Isi25] E. Ising, *Beitrag zur Theorie des Ferromagnetismus*, Z. Phys. **31** (1925) 253.

[Iwa92] J. Iwasaki and C. Rovelli, *Gravitons as embroidery on the weave*, Int. J. Mod. Phys. D **1** (1993) 533.

[Jag00] R. Jaganathan, *An introduction to quantum algebras and their applications* [arXiv:math-ph/0003018].

[Jan96] J. C. Jantzen, *Lectures on quantum groups*, Graduate Studies in Mathematics, **6** AMS, Providence, 1996.

[Joh80] D. L. Johnson, *Topics in the theory of group presentations*, London Math. Soc. Lec. Notes **42**, Cambridge Univ. Press, Cambridge, 1980.

[Jon83] V. F. R. Jones, *Index for subfactors*, Invent. Math. **72** (1983)1.

[Jon85] V. F. R. Jones, *A polynomial invariant for knots via von Neumann algebras*, Bull. Amer. Math. Soc. **12** (1985) 103.

[Jon87] V. F. R. Jones, *Hecke algebra representations of braid groups and link polynomials*, Ann. Math. **126** (1987) 335.

[Jon90] V. F. R. Jones, *Baxterization*, Int. J. Mod. Phys. B **4** (1990) 701.

[Jon91] V. F. R. Jones, "Subfactors and Knots", AMS, Providence, 1991.

[Kac02] V. Kac, P. Cheung, "Quantum Calculus", Springer-Verlag, Berlin, 2002.

[Kas95] C. Kassel, "Quantum Groups", Graduate Texts in Mathematics **155**, Springer-Verlag, Berlin, 1995.

[Kau871] L. H. Kauffman, *State Models and the Jones polynomial*, Topology **26** (1987) 395.

[Kau872] L. H. Kauffman, "On Knots", Annals of Mathematics Studies **115**, Princeton Univ. Press, Princeton, 1987.

[Kau88] L. H. Kauffman, *New invariants in the theory of knots*, Amer. Math. Monthly **95** (1988) 195.

[Kau89] L. H. Kauffman, *Statistical mechanics and the Jones polynomial*, AMS Contemp. Math. Series **78** (1989) 263.

[Kau91] L. H. Kauffman "Knots and Physics", World Scientific, Singapore, 1991.

[Kaw96] A. Kawauchi, "A Survey of Knot Theory", Birkhäuser, Basel, 1996.

[Ker00] R. Kerner, *Ternary algebraic structures and their applications in physics*, proceedings of the Conference ICGTMP "Group-23", Dubna, Russia, 2000 [arXiv:math-ph/0011023].

[Koh90] T. Kohno, "New Developments in the Theory of Knots", World Scientific, Singapore, 1990.

[Kon93] M. Kontsevich *Vassiliev's knot invariants*, Adv. Soviet Math. **16** (1993) 137.

[KR19] I. Kuntz and R. da Rocha, *Spacetime instability due to quantum gravity*, Eur. Phys. J. C **79** (2019) 447 [arXiv:1903.10642 [hep-th]].

[KT08] C. Kassel, V. Turaev, *Braid groups*, Graduate Texts in Mathematics **247**, Springer-Verlag, Berlin, 2008.

[Kul90] P. P. Kulish (Ed.), *Quantum groups*, Lecture Notes in Mathematics **1510**, Springer, Berlin, 1990.

[Lan69] L. D. Landau, E. M. Lifshitz, "Mechanics" Vol. 1, Elsevier, Oxford, 1969.

[Lis48] J. B. Listing, "Vorstudien zur Topologie", Göttingen Studien, University of Göttingen, Germany, 1848.

[Liv93] C. Livingston, *Knot theory*, The Carus Mathematical Monographs **24**, AMS, Providence, 1993.

[LK06] T. Liko and L. H. Kauffman, *Knot theory and a physical state of quantum gravity*, Class. Quant. Grav. **23** (2006) R63 [arXiv:hep-th/0505069 [hep-th]].

[LM71] M. G. G. Laidlaw, C. DeWitt-Morette, *Feynman functional integrals for systems of indistinguishable particles*, Phys. Rev. **D3** (1971) 1375.

[LM77] J. M. Leinaas, J. Myrheim, *On the theory of identical particles*, Nuovo Cimento **37B** (1977) 1.

[LR97] W. B. R. Lickorish, *An introduction to knot theory*, Graduate Texts in Mathematics **175**, Springer-Verlag, New York, 1997.

[Maj93] S. Majid, *Free braided differential calculus, braided binomial theorem and the braided exponential map*, J. Math. Phys. **34** (1993) 4843.

[Maj95] S. Majid, *Solutions of the Yang-Baxter equations from Braided-Lie algebras and braided groups*, J. Knot Theor. Ramifications **4** (1995) 673.

[Mar35] A. A. Markov, *Über die freie Aquivalenz der geschlossen Zopfe*, Recueil Math. Moscov **1** (1935) 73.

[McG64] J. B. McGuire, *Study of exactly soluble one-dimensional N-body problems*, J. Math. Phys. **5** (1964) 622.

[MD14] M. Dehn, *Die beiden Kleeblattschlingen*, Math. Ann. **75** (1914) 402.

[MMN79] C. DeWitt-Morette, A. Masheshwari, B. Nelson, *Path integration in non-relativistic quantum mechanics*, Phys. Rep. **50** (1979) 255.

[Mor72] C. DeWitt-Morette, *Feynman's path integral. Definition without limiting procedure*, Comm. Math. Phys. **28** (1972) 47.

[MT92] M. Ludde and F. Toppan, *Matrix solutions of Artin's braid relations*, Phys. Lett. B **288** (1992) 321.

[Mur87] K. Murasugi, *Jones polynomials and classical conjectures in knot theory*, Topology **26** (1987) 187.

[Nak90] M. Nakahara, "Geometry, Topology and Physics", Institute of Physics Publ., Bristol, 1990.

[Oku931] S. Okubo, *Triple products and Yang-Baxter equation. 1. Octonionic and quaternionic triple systems*, J. Math. Phys. **34** (1993) 3273.

[Oku932] S. Okubo, *Triple products and Yang-Baxter equation. 2. Orthogonal and symplectic ternary systems*, J. Math. Phys. **34** (1993) 3292.

[Ons44] L. Onsager, *Crystal statistics. I. A two-dimensional model with an order-disorder transition*, Phys. Rev. Series II, **65** (1944) 117.

[Pat72] R. K. Pathria, "Statistical Mechanics", Pergamon, Oxford, 1972.

[PR15] R. Paszko and R. Rocha, *Quadratic gravity from BF theory in two and three dimensions*, Gen. Rel. Grav. **47** (2015) 94.

[PS86] A. Pressley, G. Segal, "Loop Groups", Oxford Univ. Press, Oxford, 1986.

[RAK15] J. Roberts, M. Akveld, K. Kusejko, *Introduction to Knot Theory FS 2015*, ETH 2015.

[RBH11] R. da Rocha, A. E. Bernardini and J. M. Hoff da Silva, Exotic Dark Spinor Fields, JHEP **04** (2011) 110 [arXiv:1103.4759 [hep-th]].

[RBV10] R. da Rocha, A. E. Bernardini and J. Vaz, Jr., *k-deformed Poincare algebras and quantum Clifford-Hopf algebras*, Int. J. Geom. Meth. Mod. Phys. **7** (2010) 821 [arXiv:0801.4647 [math-ph]].

[Rei27] K. Reidemeister, *Elementare Begründung der Knotentheorie*, Abh. Math. Sem. Univ. Hamburg **5** (1927) 24.

[Rol03] D. Rolfsen, "Knots and Links", AMS, Providence, 2003.

[Roc21] R. da Rocha, *Information entropy in AdS/QCD: Mass spectroscopy of isovector mesons*, Phys. Rev. D **103** (2021) 106027 [arXiv:2103.03924 [hep-ph]].

[RH08] R. da Rocha and J. M. Hoff da Silva, *ELKO, flag-pole and flag-dipole spinor fields, and the instanton Hopf fibration*, Adv. Appl. Clifford Algebras **20** (2010) 847 [arXiv:0811.2717 [math-ph]].

[Rov88] C. Rovelli and L. Smolin, *Knot theory and quantum gravity*, Phys. Rev. Lett. **61** (1988) 1155.

[RT20] R. da Rocha and A. A. Tomaz, *Hearing the shape of inequivalent spin structures and exotic Dirac operators*, J. Phys. A **53** (2020) 465201 [arXiv:2003.03619 [hep-th]].

[RV07] G. Rote, G. Vegter, *Computational Topology: An Introduction*, in "Effective Computational Geometry for Curves and Surfaces", (J. Boissonnat, M. Teillaud, eds.), Springer-Verlag, Berlin, 2007.

[Sch68] L. Schulman, *A path integral for spin*, Phys. Rev. **176** (1968) 1558.

[Sch70] J. Schwinger, *Unitary operator bases*, Proc. Nat. Acad. Sci. **46** (1960) 570; in "Quantum Kinematics and Dynamics", Benjamin, New York, 1970.

[Sum95] Sumners, *Lifting the curtain: Using topology to probe the hidden action of enzymes*, Notices of the AMS **42** (1995) 528.

[Thi87] M. Thistlethwaite, *A spanning tree expansion of the Jones polynomial*, Topology **26** (1987) 297.

[Tho82] J. J. Thomson, *On the vibrations of a vortex ring, and the action of two vortex rings upon each other*, Phil. Trans. R. Soc. London, Ser. A **173** (1882) 493.

[Tho83] J. J. Thomson, "A Treatise on the Motion of Vortex Rings", Macmillan, London, 1883.

[Tho85] J. J. Thomson and H. F. Newall, *On the formation of vortex rings by drops falling into liquids, and some allied phenomena*, Proc. R. Soc. London, Ser. A **39** (1885) 417.

[TL71] N. Temperley, E. Lieb, *Relations between the percolation and colouring problem and other graph-theoretical problems associated with regular planar lattices: some exact results for the percolation problem*, Proc. Royal Soc. Ser. A **322** (1971) 251.

[Tur94] V. Turaev, *Quantum invariants of knots and three manifolds*, De Gruyter Stud. Math. **18** (1994) 1-588.

[Van71] A. T. Vandermonde, *Remarques sur les problèmes de situation*, Memoires de l'Académie Royale des Sciences (Paris) (1771) 566.

[Vas90] V. A. Vassiliev, *Cohomology of knot spaces*, "Theory of singularities and its applications", 23-69, Adv. Soviet Math. **1**, AMS, Providence, 1990.

[VR16] J. Vaz, Jr. and R. da Rocha, "An Introduction to Clifford Algebras and Spinors," Oxford Univ. Press, Oxford, 2016.

[VRH95] S. Viefers, F. Ravndal, T. Haugset, *Ideal quantum gases in two dimensions*, Am. J. Phys. **63** (1995) 69.

[WDC85] S. Wasserman, J. Dungan, N. Cozzarelli, *Discovery of a predicted DNA knot substantiates a model for site-specific recombination*, Science **229** (1985) 171.

[Wen88] H. Wenzl, *Hecke algebras of type A_n and subfactors*, Invent. Math. **92** (1988) 349.

[Wit89] E. Witten, *Quantum field theory and the Jones polynomial*, Comm. Math. Phys. **121** (1989) 351.

[Wol18] Wolfram Research, Inc., Mathematica, Version 11.3.0.0, Champaign, IL (2018)
https : //mathworld.wolfram.com/SolomonsSealKnot.html

[Wu92] F. Y. Wu, *Knot theory and statistical mechanics*, Rev. Mod. Phys. **64** (1992) 1099.

[YG89] C. N. Yang, M. L. Ge, "Braid Group, Knot Theory and Statistical Mechanics", World Scientific, Singapore, 1989.

[ZW00] Z. Zhang, J. Wang, *Some properties of the (q, h)-binomial coefficients*, J. Phys. A: Math. Gen. **33** (2000) 7653.

Index

Printed in the United States
by Baker & Taylor Publisher Services